PRACTICAL MANUAL FOR BOTANY

Vol. I

PRACTICAL MANUAL FOR BOTANY

Vol. I

Elizabeth Margaret
READER AND HEAD

G. Angela
SENIOR FACULTY

Department of Botany
St. Anne's Degree College for Women,
Mehidipatnam, Hyderabad-500028

PUBLISHING FOR ONE WORLD

NEW AGE INTERNATIONAL (P) LIMITED, PUBLISHERS
(formerly Wiley Eastern Ltd.)

New Delhi • Bangalore • Chennai • Cochin • Guwahati • Hyderabad
Jalandhar • Kolkata • Lucknow • Mumbai • Ranchi
Visit us at **www.newagepublishers.com**

Published by New Age International (P) Ltd., Publishers
First Edition: 2008

ISBN (10): 81-224-2216-0

ISBN (13): 978-81-224-2216-0

C-07-09-2029

Printed in India at Ramprintograph, Delhi.
Typeset at Aakriti Graphics, Delhi.

PUBLISHING FOR ONE WORLD

NEW AGE INTERNATIONAL (P) LIMITED, PUBLISHERS
(formerly Wiley Eastern Limited)
4835/24, Ansari Road, Daryaganj, New Delhi - 110002
Visit us at **www.newagepublishers.com**

Preface

This textbook "Practical Manual for Botany. Vol-I" has been written in accordance with the revised common core syllabus for the first year undergraduate students.

This edition comprises four branches: Algae, Fungi, Bryophytes and Pteritophytes. Each type has been presented in a simple and concise manner with appropriate diagrams. The contents of the book are perfectly fine tuned to the requirements of target segment. It endeavors to meet the vital need of creating in the minds of I year degree students — a genuine interest for the subject and also a healthy appetite for learning. It is proposed to follow up this publication shortly — with practical manuals on the same lines for II and III year students.

Authors

Botany Laboratory: General Instructions to Students

The following laboratory rules are to be observed at all times by a student:

1. A student must attend to laboratory work with utmost sincerity, honesty and inquisitiveness.
2. Assigned work to be completed within specified time.
3. Careful handling of reagents and microscopes.
4. At the end of the laboratory period, working space should be left clean and in order.
5. Work strictly according to the instructions of the teacher.
6. Work in the laboratory strictly on individual basis, unless and until indicated specifically for teamwork.
7. Maintain an observation notebook and a record.
8. To carry necessary instruments and stationary items including slides, cover slips, extra record sheets to every practical class.
9. Every student should keep herself busy with own work and whenever necessary consult the teacher alone.
10. Students should take care of laboratory cleanliness while working and wash their hands with soap and water before leaving the laboratory.

Compound Microscope

Description of a Microscope

1. The microscope consists of a strong basal foot and an arm. A stage which is rectangular or round is fixed to the lower end of the arm.
2. The microscope has two lens : an objective and an eye pieces (occular).
3. Objective lens produces a primary image of the material and eye piece produces a magnified image of the primary image. This microscope is also called monocular monobjective microscope.
4. The body of microscope consists of a tube with an occular attached at the upper end and a revolving nose-piece with third objetive at the other end. The three objectives are low power (10x), high power (40x) and oil immersion (100x).

   The diaphragm is a wheel shaped metal disc to regulate the aperture, through which light rays reach the condenser and are passed to the object. Condenser is a septum with two or more lenses which receive parallel light from the mirror and converge them at the level of the stage.
5. A concave mirror is fixed at the lower part of the limb to focus the converging cone of rays at the level of specimen.
6. The tube of microscope is movable with the help of coarse and fine adjustments screw on the limb.

Steps to Handle Microscope

1. Clean the mirror and lenses before and after uses with soft muslin cloth or tissue paper.
2. Turn thc mirror towards the light and view through the eye piece.
3. Light is adjusted by closing and opening the diaphragm.
4. Place the slide on the stage and fix it with the clamps.
5. First view through the low power by adjusting the coarse adjustments (Keep both the eyes open while viewing)
6. To view under higher magnification, first raise the body upwards, rotate the nose-piece to high power. Then use the fine adjustment.
7. Don't move the slide once the fine adjustments are made.

After using the microscope, clean it with muslin cloth and keep it back in the cabinet.

Compound Microscope

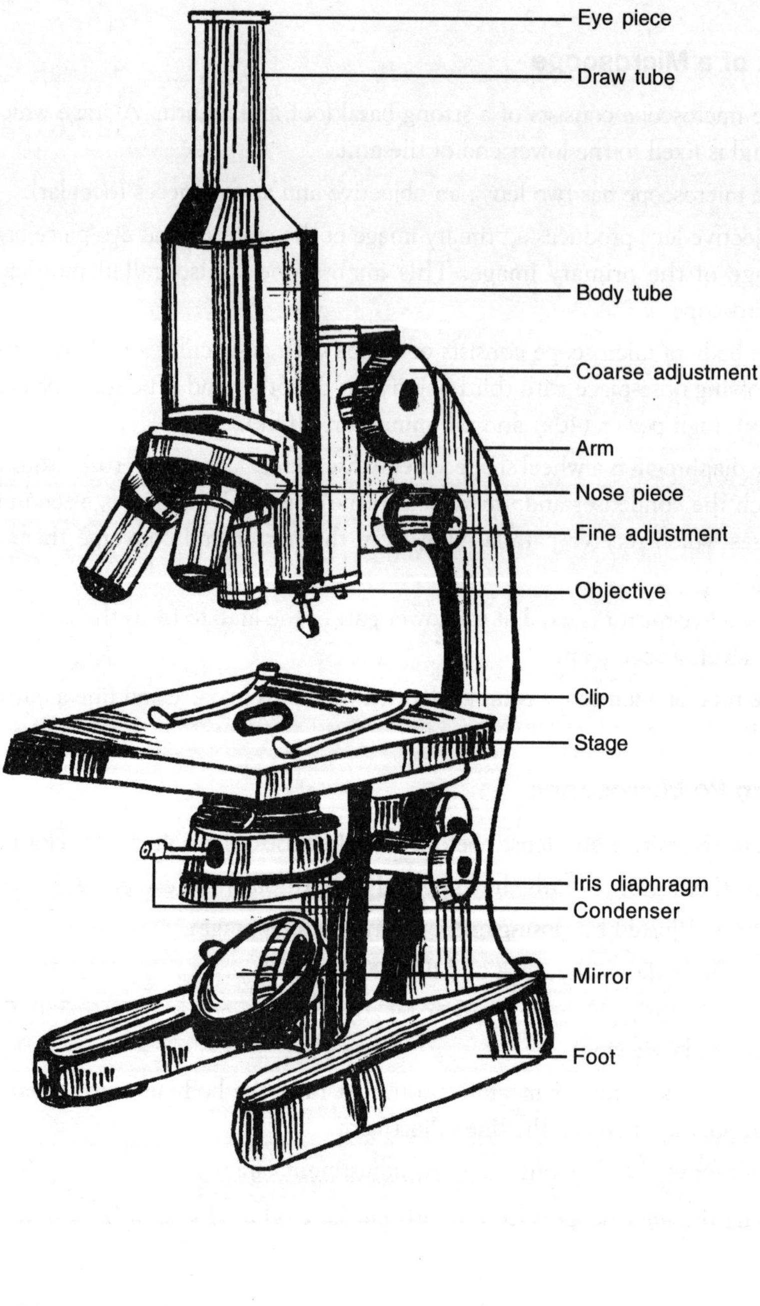

Contents

1 CHAPTER

ALGAE

1.1 INTRODUCTION

The word 'algae' is derived from a Latin word 'alga' which means 'Sea weed'. Algae are chlorophyll bearing autotrophic thallophytes. Study of algae is called Phycology (Greek word 'phycos' means sea weed and 'logos' means study). This group is one of the largest of the Plant Kingdom.

Algae are distributed in almost all the types of habitats. Plants are thalloid and show great variation in thallus organisation. They include motile unicellular forms, motile colonial forms, palmelloid, filamentous, heterotrichous, siphonaceous, uniaxial and multiaxial thalli.

The cells constituting the thalli are basically of two kinds prokaryotic and eukaryotic The cells in all members of algae remains surrounded by a cellulose cell wall which encloses the protoplast. Members belonging to Cyanophyceae have a prokaryotic cellular organisation while all the other members show eukaryotic organisation.

Plants reproduce by vegetative, asexual and sexual methods. In Cyanophyceae sexual reproduction is not known. Vegetative reproduction is by fragmentation, fission, akinetes, bulbils, amorphous bulbils, amylum stars, hormogones or by tubers. Asexual reproduction is by formation of zoospores, aplanospores, hypnospores and tetraspores. Sexual method of reproduction ranges from isogamy to oogamy.

Algae have been variously classified by numerous phycologists. The most simple and practical classification was proposed by F.E. Fritsch (1935, 1945). Fritsch classified algae into 11 classes. They are

1. Chlorophyceae
2. Xanthophyceae
3. Chrysophyceae
4. Bacillariophyceae
5. Cryptophyceae
6. Dinophyceae
7. Chloromonadinae
8. Euglenineae
9. Phaeophyceae
10. Rhodophyceae
11. Myxophyceae (Cyanophyceae)

1.2 OSCILLATORIA

Division — Cyanophyta
Class — Cyanophyceae
Order — Nostocales
Family — Oscillatoriaceae
Genus — *Oscillatoria*

Identifying Characters

1. Trichomes are without heterocysts and without branching (A single row of cells in a filamentous colony is known as trichome). The trichome along with sheath is called filament.
2. Biconcave separating discs are present.
3. Free surface of the terminal cell is thickened to form calyptra.

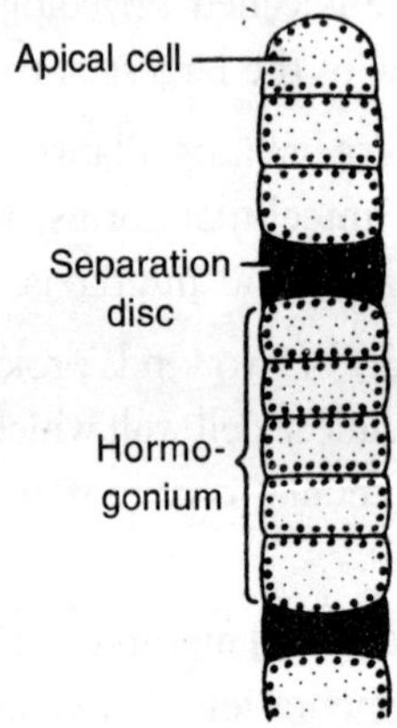

Fig. 1.1 Oscillatoria : Thallus structure

Distinguishing Characters

1. Majority of the species inhabit fresh water while others are marine. They are commonly found as floating masses in rainwater, pools, puddles and also found on moist soils.
2. They may occur singly or interwoven with each other forming extensive masses on terrestrial substrate.
3. It is an unbranched filamentous alga.
4. Each trichome is comprised of numerous cells, which are broader than long and are placed end to end in a row. They are unbranched, cylindrical, lacks a sheath or possess delicate inconspicuous sheath.
5. All cells resemble each other except the terminal cell, which is colourless and variously shaped. This cell has thick outer covering called calyptra.

6. Cells are prokaryotic. Each cell has peripheral coloured chromoplasm and central colourless centroplasm. Pigments and reserve food materials are found in chromoplasm. In centroplasm nuclear material is seen.
7. In floating species pseudovacuoles (gas vacuole) are present in the chromoplasm.
8. Trichome exhibit gliding, rotatory, oscillatory movements hence the name *Oscillatoria.*
9. Reproduction is by fragmentation and hormogone formation.
10. The broken filaments which consist of 2–3 cells are called hormogones.
11. Hormogones are formed due to death of intercalary cells or by special separating biconcave cells or disc.

1.3 NOSTOC

Division — Cyanophyta
Class — Cyanophyceae
Order — Nostocales
Family — Nostocaceae
Genus — *Nostoc*

Identifying Characters

1. Trichomes unbranched, moniliform and contorted.
2. Akinetes and intercalary heterocysts are present.
3. Algae form jelly like balls in which numerous filaments are embedded.
4. Cells are prokaryotic.

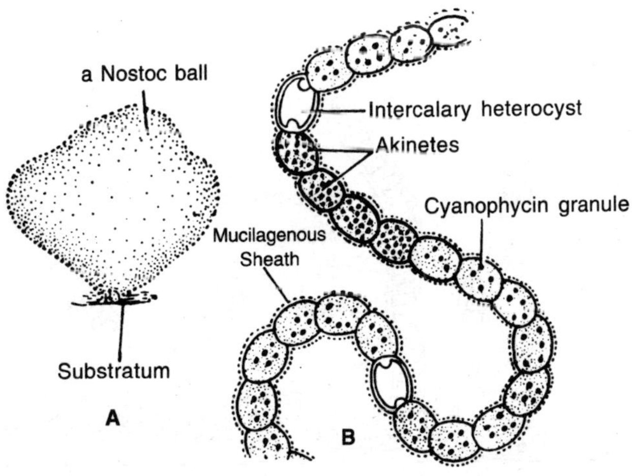

Fig. 1.2 Nostoc: A. Gelatinous colony, B. Trichome

Distinguishing Characters

1. They occur both in aquatic and terrestial habitats.
2. Thallus is filamentous. Filaments do not occur singly but often grow as large colonies of closely packed trichomes, embedded in a firm matrix of gelatinous material.
3. Each trichome is enclosed by its own mucilaginous sheath and is called a filament.
4. Colonies are spherical in shape, solid or hollow. Surface may be smooth or verrucose.
5. Trichomes are unbranched, contorted and moniliform.
6. Cells of the trichome are spherical or oval and joined end to end resembling a string of beads.
7. In between the vegetative cells there are specialized enlarged thick walled empty looking cells called heterocysts.
8. Cells are prokaryotic. They have centrally located centroplasm and peripheral coloured chromoplasm having pigments and reserve food materials.
9. Heterocysts are intercalary or terminal. They are spherical, enlarged, thick walled yellow coloured cells. Each heterocyst has two polar pores plugged by polar nodules, each one adjacent to the neighbouring cell. Terminal heterocysts have a single polar nodule. These help in nitrogen fixation and hormogone formation.
10. Reproduction is by the formation of hormogones and akinetes. Sexual reproduction is absent.
11. Akinetes are thick walled resting resistant spore like vegetative cells. These occur in chains next to heterocyst.

1.4 SCYTONEMA

Division — Cyanophyta
Class — Cyanophyceae
Order — Nostocales
Family — Scytonemataceae
Genus — *Scytonema*

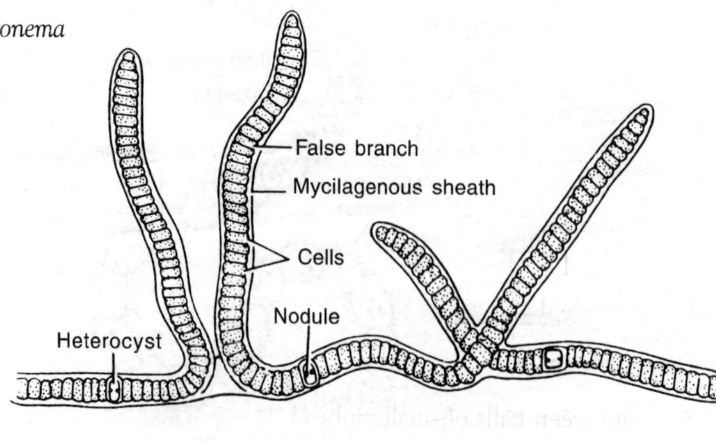

Fig. 1.3 Scytonema : Thallus structure

Identifying Characters

1. Trichomes show false branching.
2. Trichomes have heterocysts.

Distinguishing Characters

1. Commonly occurs in sub-aerial habitats. Some are free floating in aquatic habitats.
2. Filaments occur single and trichome remains enveloped in sheath.
3. Filaments show false branching, which is single or geminate (in pairs).
4. False branching occurs due to breakage of trichome within the sheath and one or both the broken ends of trichome protrudes out of the sheath and develop into branches.
5. False branching occurs due to degeneration of one or more intercalary cells by the breakage of trichome near the intercalary heterocyst or by the formation of loops.
6. Trichomes are of same diameter throughout with cylindrical cells. Cells are shorter than broad.
7. Trichome is covered by an individual sheath, which is firm and hyaline or coloured.
8. Heterocysts are intercalary and are of the same size as that of vegetative cells. Branching occurs at the region of hererocyst.
9. Heterocysts are thick walled specialized vegetative cells, help in nitrogen fixation.
10. Cells are prokaryotic with outer coloured chromoplasm and inner colourless centroplasm.
11. Sexual reproduction is absent.
12. Hormogone formation is common mode of multiplication. They are formed at the ends of trichome.

1.5 VOLVOX—COLONY

Division — Chlorophyta
Class — Chlorophyceae
Order — Volvocales
Family — Volvocaceae
Genus — *Volvox*

Identifying Characters

1. Thallus is a coenobium.
2. Coenobia are spherical or oval.
3. Chloroplast of individual cell is cup shaped.

Distinguishing Characters

1. It occurs as minute green balls of small pinhead size, in temporary and permanent fresh water pools and ponds.

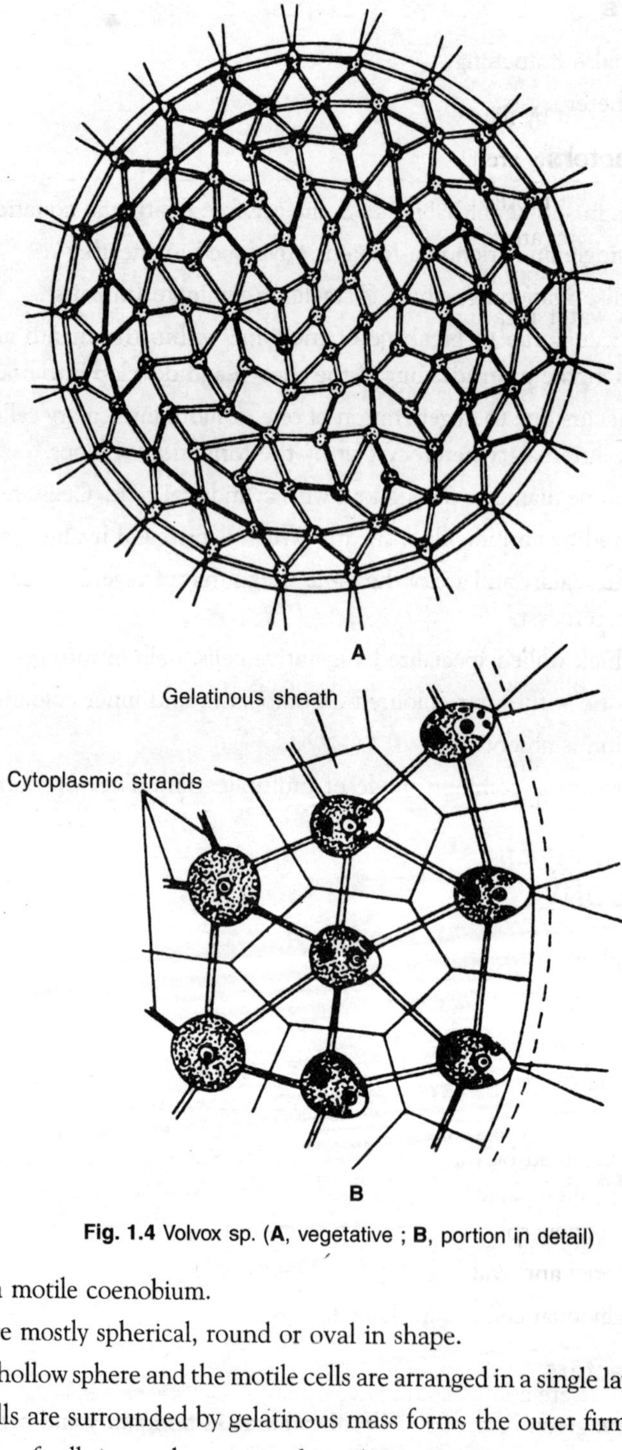

Fig. 1.4 Volvox sp. (**A**, vegetative ; **B**, portion in detail)

2. Thallus is a motile coenobium.
3. Colonies are mostly spherical, round or oval in shape.
4. Colony is a hollow sphere and the motile cells are arranged in a single layer towards the periphery.
5. Layer of cells are surrounded by gelatinous mass forms the outer firm limiting layer.
6. The number of cells in a colony varies from 500 – 60,000 according to species.

7. Each cell in the colony is connected with a few of the neighbouring cells by thin and delicate cytoplasmic strands.
8. Each cell is surrounded by its own gelatinous sheath.
9. All the cells in the colony are *Chlamydomonas* like.
10. Each cell is biflagellate motile and ovoid. Flagella are whiplash type and anteriorly inserted. Two contractile vacuoles are situated at the base of flagella. Cytoplasm has cup shaped chloroplast with a pyrenoid, an eye spot and a single nucleus. Cytoplasm is rich in volutin granules.
11. It moves through water rotating slowly with one end of the sphere always leading - hence rolling alga.
12. Cells in the colony show division of labour - only few cells take part in reproduction.
13. Asexual reproduction is by formation of daughter colonies.
14. Sexual reproduction is Oogamous.
15. Colonies may be monoecious or dioecious.
16. Male reproductive organs are antheridia, which develop at the posterior part of the colony.
17. Each antheridium produces 128 or 512 biflagellate sperms, which are liberated as bowl shaped plate or hollow sphere.
18. Female sex organ is oogonium formed at the posterior end. It produces a single egg which is non-motile.
19. Zygote produced as a result of fertilization is thick walled and orange red in colour. The outer layer is often ornamented.

VOLVOX—Asexual Reproduction by Daughter Colonies

1. Asexual reproduction in *Volvox* is by means of daughter colonies.
2. *Volvox* is a motile colonial coenobium, found in fresh waters.
3. The individual cells are Chlamydomonas like and arranged at the periphery of the coenobium.
4. Asexual reproduction takes place by formation of daughter colonies.
5. About 2 – 50 cells in the posterior half of the colony enlarge in size and are called gonidial cells, which by repeated longitudinal divisions give rise to daughter colonies.
6. The eight cells are formed by the longitudinal division of the gonidial cells are arranged in a curved plate and is called plakea stage, which by further longitudinal divisions produce several cells arranged in a hollow sphere with a small aperture called phialopore.
7. All the cells are naked and without flagella with anterior end pointed towards the center.
8. Normal orientation is attained by complete inversion of the young coenobium through phialopore.
9. Naked protoplasts secrete cell wall and develop flagella at their anterior ends and develop into daughter colony.
10. The young colonies escape by the disintegration of the parent colony or through a pore at the position of the original gonidium.

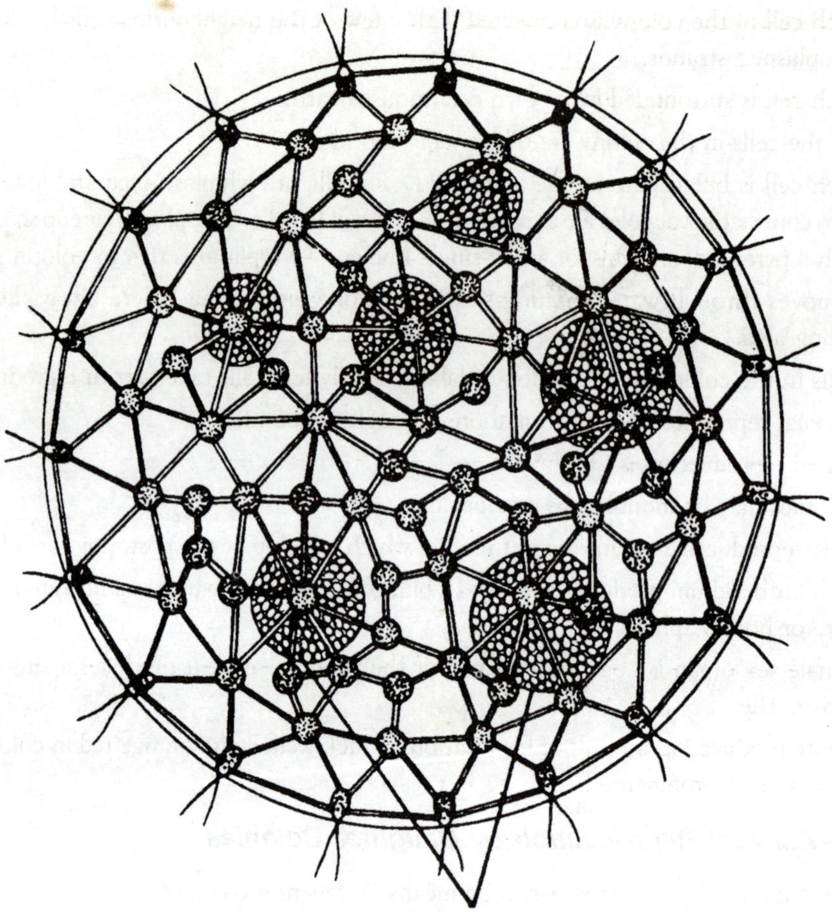

Fig. 1.5 Volvox sp. parent colony with a number of daughter colonies

VOLVOX—Male Colony Showing Antheridia

1. Sexual reproduction in *Volvox* is oogamous.
2. Antheridia are male reproductive organs.
3. Sex organs may be monoecious or dioecious.
4. Certain cells in the posterior region of the colony enlarge, retract their flagella and form gametangia.
5. The male gametangia are called the antheridia.
6. The protoplast of the antheridium divides longitudinally several times and produces 64 - 128 cells, which after inversion are grouped as bowl shaped plate.
7. In some species the divisions continue and produce 512 cells arranged in the form of the hollow sphere. Each cell develops into a spermatozoid or sperm.

8. Each sperm is biflagellate conical or fusiform structure.
9. The entire mass of biflagellate sperms are liberated as a single unit, which swims freely until it reaches the female colony.

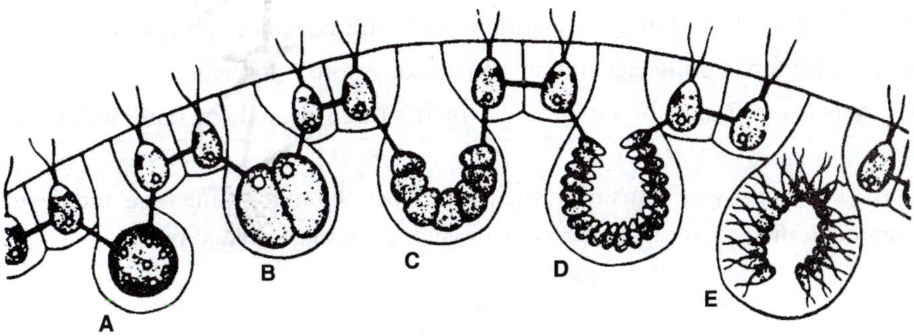

Fig. 1.6 Volvox colony showing Artheridum A. Antheridium B-C development of sperms D-E antherozoids before and after inversion

VOLVOX—Showing Oogonia—Female Colony

1. Oogonia are female reproductive structures.
2. *Volvox* is a motile coenobial form. It reproduces sexually by oogamous method.
3. Sex organs may be monoecious or dioecious.
4. Few cells at the posterior region enlarge, become rounded or flask shaped, retract their flagella and develop into oogonia.
5. The entire protoplast of the oogonium gets metamorphosed into a single egg or oosphere.
6. The egg has a large central nucleus and a parietal chloroplast having a large central nucleus and numerous pyrenoids and also stored with reserve substances.
7. The oosphere develops a beak like protrusion, which marks the point of entrance of sperm.

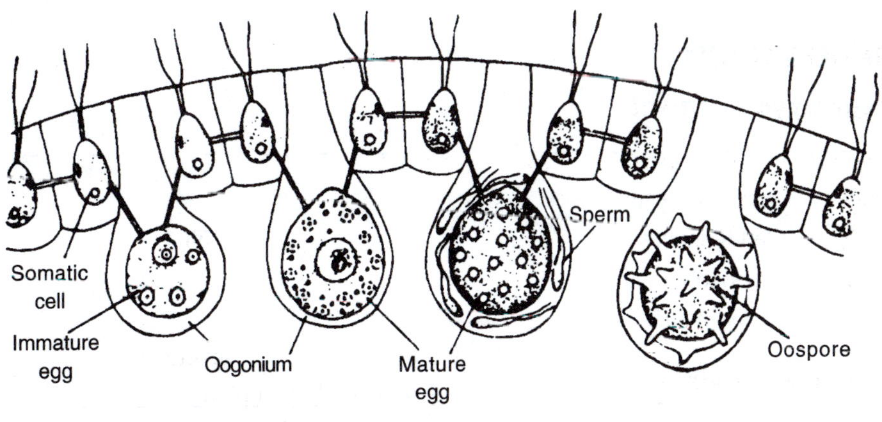

Fig. 1.7 Volvox colony showing oogonium and oospore

VOLVOX—Zygote

1. In *Volvox* sexual reproduction is oogamous, and colonies may be monoecious or dioecious.
2. Antheridia and oogonia are developed mostly in the posterior part of the colony.
3. Antheridium produces biflagellate antherozoids and oogonium produces a single egg.
4. Antherozoids fertilize the egg and an oospore or zygote is formed.
5. The zygote is surrounded by a thick wall which is three layered and outer wall layer is smooth or spiny.
6. The thick walled zygote remains in the parent coenobium for some time and develop sufficient haematochrome or red pigment to colour their protoplast orange red.

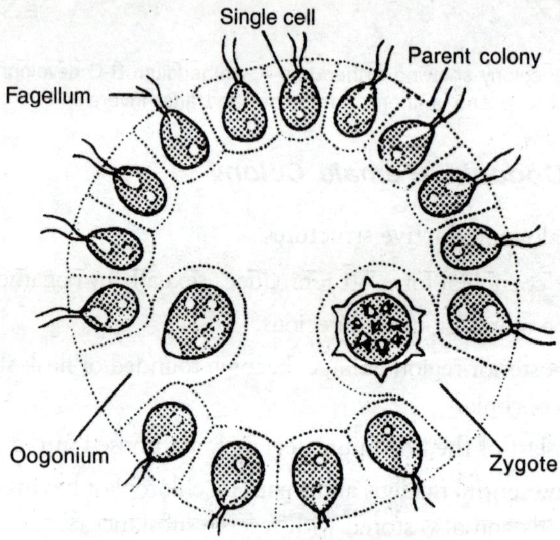

Fig. 1.8 Volvox. A colony with a zygote

1.6 OEDOGONIUM

Vegetative Filament

Division — Chlorophyta
Class — Chlorophyceae
Order — Oedogoniales
Family — Oedogoniaceae
Genus — *Oedogonium*

Identifying Characters

1. Thallus multicellular, filamentous and unbranched.
2. Presence of cap cells.

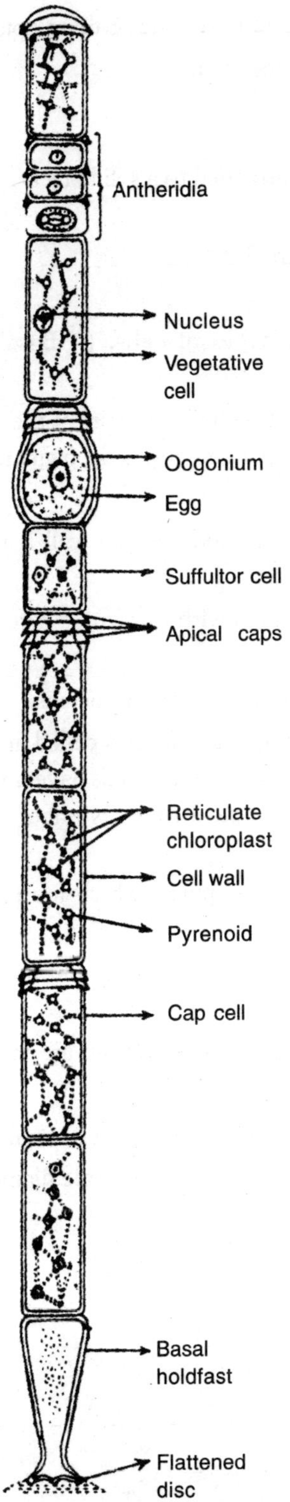

Fig. 1.9 Oedogonium. A monoecious filament

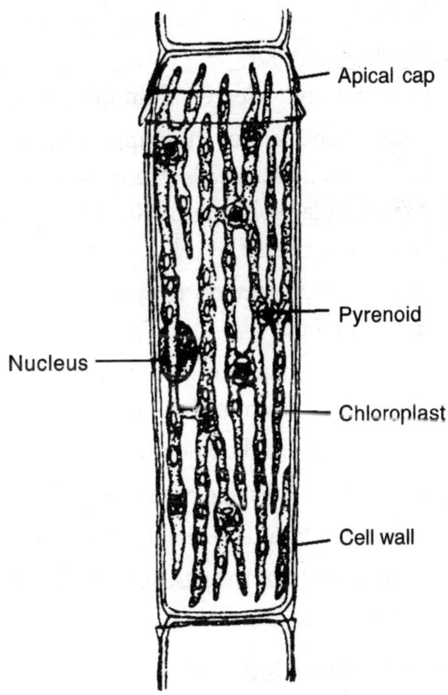

Fig. 1.10 Oedogonium cell

3. Oogonia spherical, larger than the vegetative cell and bears one or more apical rings.
4. Chloroplast is reticulate with pyrenoids at the intersections.

OEDOGONIUM–Vegetative Structure

1. It is a common submerged aquatic alga often found attached to a solid object. Mature filaments are free floating.
2. Thallus is filamentous, multi-cellular and unbranched.
3. Filaments are uniseriate.
4. According to position the filaments have three types of cells, (*i*) Basal (*ii*) Intercalary (*iii*) Apical.
5. The basal cell of the filament functions as a hold fast. The lower part of the hold fast is either disc shaped or finger like. Upper part is broad and rounded and lacks pigments.
6. The terminal or apical cell is broadly rounded.
7. The intercalary cells are cylindrical longer than broad and the cell wall is made up of three layers inner cellulose, middle pectin and outer chitin.
8. Internal to cell wall the cells have reticulate chlorplasts with numerous pyrenoids at the inter sections.
9. Cells are uninucleate. Thin delicate cytoplasmic strands suspend nucleus.
10. Some of the cells at their upper end possess small ring like markings called apical rings. These cells are called cap cells. Presence of apical rings indicates that the cells have undergone division. The annular splitting of the lateral cell walls forms apical rings. Number of rings indicates number of times the cell has divided.
11. Asexual reproduction is by means of multiflagellate zoospores, which are produced singly in zoosporangia.
12. Sexual reproduction is oogamous. Female sex organ is oogonium and male sex organs are antheridia. Each antheridium produces two multi-flagellate sperms.

    On the basis of distribution of antheridia two species of Oedogonium are identified.

    (1) Macrandrous and (2) Nannandrous.
14. In Macrandrous species, the antheridia are found on the filaments of normal size.
15. In Nannandrous species, antheridia are produced on much reduced male filaments called dwarf males.

OEDOGONIUM—Cap Cells

1. *Oedogonium* is a fresh water alga found in pools, lakes, tanks, etc.
2. The thallus is multicellular, filamentous and unbranched.
3. The basal cells of the filament function as hold fast.
4. The cell at the tip of the filament is the apical cell.

5. The intercalary cells are cylindrical with thick walls composed of 3 layers.
6. Certain cells in every filament possess one or more ring like markings known as apical caps at their distal ends and these cells are known as cap cells characteristic feature of the *Oedogonium* species.
7. The caps are formed as a result of the cell division. The number of caps in a cell indicates the number of times a cell has divided. When a cell divides a thickening is laid down in its upper end in the form of a ring on the inner side of the cell. The latter now stretches to form a new cell. Simultaneously the nucleus of the cell moves upward, divides mitotically into two and one of them passes into the newly formed cell, while the other remains in the parent cell. The new cell therefore has the wall chiefly formed by the stretching of the thickened wall material while the cap in fact is a small portion of the old cell wall.

OODOGONIUM—Vegetative with Hold Fast

1. *Oedogonium* is a fresh water green alga found in pools, lakes, tanks, etc.
2. The thallus is multicellular, filamentous and unbranched.
3. The basal cells of the filament function as a hold fast.
4. The hold fast is generally colourless but may contain well-developed chloroplast and produce certain out growths, which help in attachment of the filament to the substratum.
5. The cell at the tip of the filament is apical cell. The intercalary cells are cylindrical with thick wall composed of three layers.
6. Certain cells in every filament possess one or more ring like markings known as apical caps at their distal ends and these cells are known as cap cells characteristic of Oedogonium species.
7. Reticulate chloroplast lies internal to cell wall with many pyrenoids.
8. The cells are uninucleate. Cytoplasmic strands suspend the single nucleus.

OEDOGONIUM—Macrandrous Male

1. In Macrandrous species, antheridia are produced on the filaments of normal size.
2. The sexual reproduction in *Oedogonium* is oogamous and reproductive structures are oogonia and antheridia.
3. In some species the antheridia and oogonia are produced on two different filaments. Such species are heterothallic and are called dioecious species.
4. The filaments bearing antheridia are called macrandrous male. Macrandrous means the male and female filaments are of the same size.
5. The antheridia are intercalary as well as terminal. They are formed from the antheridial mother cells. Each antheridial mother cell divides into two, the upper antheridial cell and lower sister cell. The antheridial cell is smaller. This cell may divide and redivide forming a chain of antheridia.
6. The antheridia are flat, short, cylindrical or disc like cells of the filament.
7. They lie in a row and their number varies from 2 to 40.

8. The contents of each antheridium develop into two sperms, rarely into one.
9. The sperms are liberated from the antheridia and each antherozoid is a pyriform structure with a colourless beak and a sub-apical crown of flagella (Stephanokont).

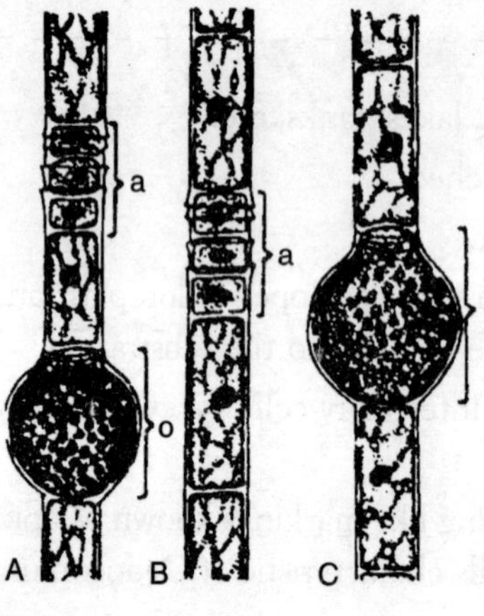

Fig. 1.11 Oedogonium. Distribution of sex organs in macrandrous sp. A, Macrandrous monoecious with the antheridia (*a*) and oogonia (*o*) on the same filament; B-C, Macrandrous dioecious with antheredia (*a*) on filament B and oogonia on filament C

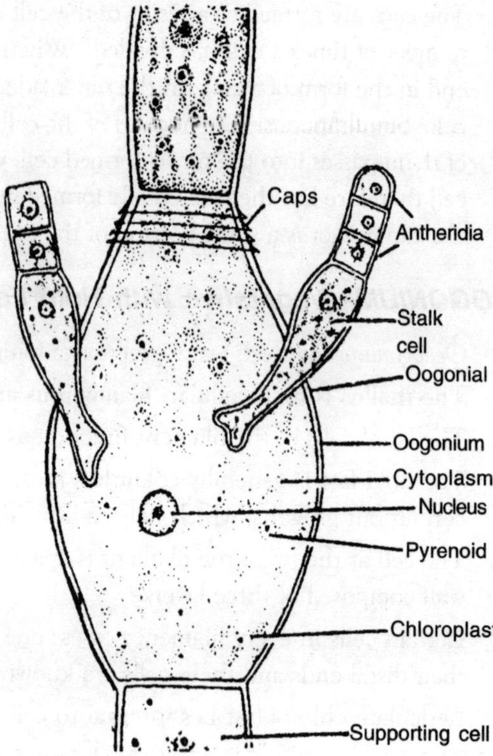

Fig. 1.12 Oedogonium. Oogonium with two dwarf males attached to it (Nannandrous species)

OEDOGONIUM—Macrandrous Female

1. The sexual reproduction in *Oedogonium* is oogamous and reproductive structures are oogonia and antheridia.
2. In Macrandrous species, the antheridia are produced on the filaments of normal size.
3. In some species, antheridia and oogonia are produced on two different filaments, such species are heterothallic and are called dioecious.
4. The filaments bearing oogonia are called macrandrous female. Macrandrous means the male and female are of same size.
5. Oogonia occur in intercalary or terminal position. Oogonia may be solitary or in a row of 2–3.

6. Any vegetative cell of a filament can give rise to oogonia. The cell which gives rise to oogonia is called oogonial mother cell. It divides transversely into two. The upper cell enlarges into a flask shaped spherical oogonium. The lower cell is known as supporting cell or suffultory cell. In some the supporting cell divides subsequently and form a chain of oogonia.
7. Oogonium encloses a single large ovum. The wall of the oogonium has a small pore on one side known as receptive pore just behind the receptive spot, which is a hyaline area in the protoplast of oogonium.
8. Oogonium is uninucleate and protoplast is rich in reserve food.

OEDOGONIUM—Nannandrous Male (Dwarf Male)

1. In Nannandrous forms male filament is smaller in size and is known as a dwarf-male. The dwarfs are produced by the germination of androspore produced within androsporangia.
2. If the androsporangia and oogonia are developed on the same filament such filaments are known as gynandrosporous.
3. If they are borne on two different filaments they are called idioandrosporous species
4. The contents of each androsporangium metamorphose into a single androspore, which is similar in structure to antherozoid but its size is intermediate between zoospore and antherozoid.
5. After a short period it rests on oogonium or its stalk and germinate to give rise to dwarf male.
6. Each dwarf male has a basal attaching cell and few antheridia. Each antheridium gives rise to two antherozoids.

1.7 CHARA

Division — Chlorophyta
Class — Chlorophyceae
Order — Charales
Family — Characeae
Genus — *Chara*

Identifying Characters

1. The thallus is macroscopic, branched and multicellular. The main axis is differentiated into nodes and internodes.
2. At the node are present lateral branches of unlimited growth and branches of limited growth.
3. Both the sex organs, the male sex organ is called 'globule' and female organ "nucule" are present at one point on the node of the short lateral branch.

CHARA—Vegetative

1. The plant body is calcified and commonly called "stone worts".
2. Chara is widely distributed in submerged conditions in fresh water pools, lakes and slow flowing streams.

3. It is attached to muddy or sandy bottom of the pond or pool by means of rhizoids.
4. The main axis consists of a series of alternating nodes and internodes.
5. The internodes consist of a single undivided, elongated cylindrical cell.
6. The node is made up of a transverse layer of short cells.
7. From the nodes of the central axis arise whorls of primary laterals of limited growth and from the axils of primary laterals branches of unlimited growth arises. The primary laterals are also divided into a few nodes and internodes. From these nodes arise the secondary laterals.
8. The internodal cells in the corticated species are surrounded by a number of sheath cells arising from the nodes above and below.
9. The growth takes place by a single dome apical cell.

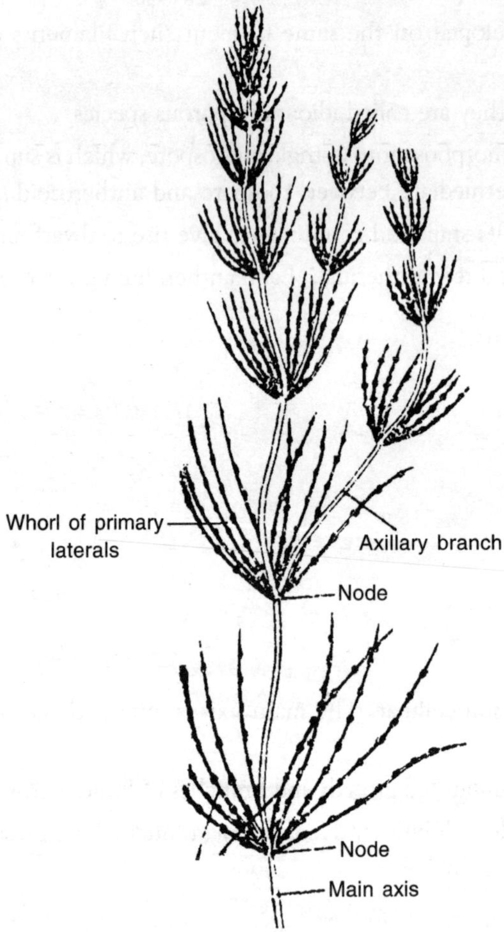

Fig. 1.13 Chara. Habit showing organisation of thallus

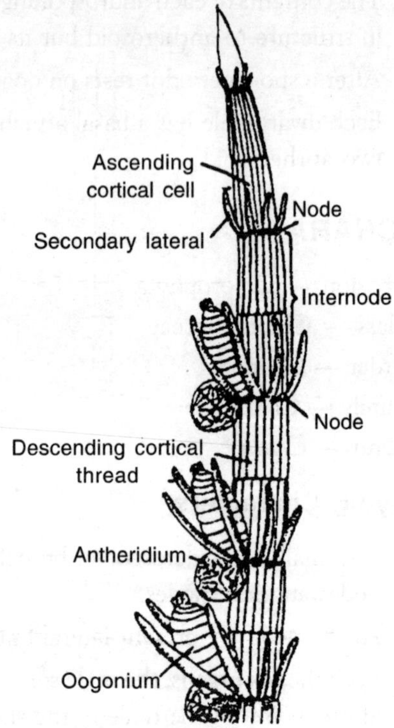

Fig. 1.14 Chara. A fertile primary lateral

CHARA—Sex Organs

1. Sexual reproduction in *Chara* is oogamous.
2. The sex organs are large, highly specialized and complicated in structure.
3. The male sex organ is a large, round bright yellow or red structure; it is globule (antheridium).
4. The female sex organ or the oogonium is a large, oval body covered with a multicellular envelope, it is known as nucule.
5. The antheridium has a wall composed of eight closely fitting large hollow curved plate like cells, the shield cells. A rod shaped cell manubrium arises from the center of the shield cell, which bears primary and secondary capitula. The secondary capitula bear branched thread like cells, the spermatogenous filaments. Small discoid cells from spermatozoid mother filaments, spirally coiled, are given off by the spermatogenous filaments. Each spermatozoid mother cell develops into a band shaped spermatozoid.
6. Nucleus is oval in shape and is situated above the globule at the node. It is enveloped by spirally coiled tube cells. At the apex of the nucule is a corona. Oosphere is single where a nucleus lies surrounded by the cytoplasm.

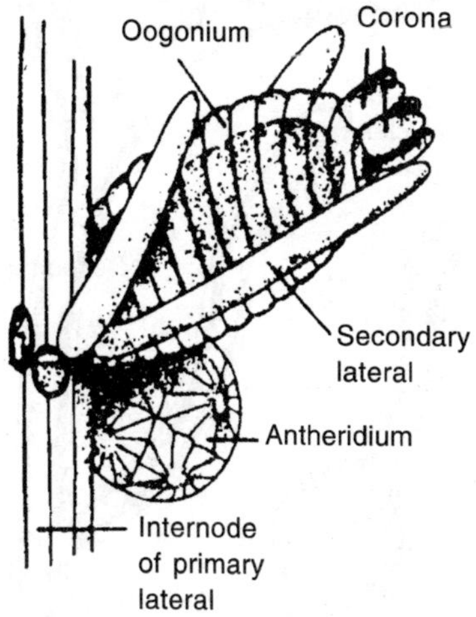

Fig. 1.15 Chara. Sex organs

1.8 VAUCHERIA—VEGETATIVE

Division — Xanthophyta
Class — Xanthophyceae
Order — Heterosiphonales
Family — Vaucheriaceae
Genus — *Vaucheria*

Identifying Characters

1. Thallus is tubular, cylindrical, branched, aseptate and coenocytic.
2. Sex organs are antheridia and oogonia.

Distinguishing Characters

1. Species of *Vaucheria* are found in both fresh and seawater.
2. Some species are terrestrial. In winter months, they occur as green matty structures on damp soils.
3. The plant body is cylindrical, aseptate, monosiphonous and branched. Branching is monopodial. However, septa are formed at the base of the reproductive organs.
4. Thallus is usually attached to the substratum by a colourless branched hold fast.

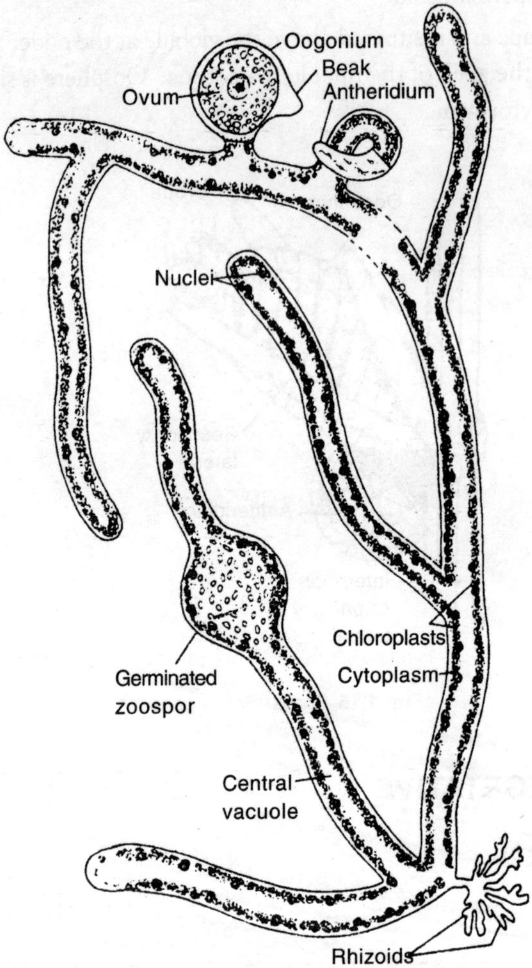

Fig. 1.16 Vaucheria sp. Filament with rhizoids

5. The plant body is multinucleate and acellular i.e., coenocytic and aseptate.
6. Cell wall is thin elastic and consists of an outer layer of pectin and an inner layer of cellulose.
7. There is a thin layer of cytoplasm, which encloses a large central vacuole that extends the entire length of the thallus.
8. Within the cytoplasm there are many chromatophores and nuclei.
9. Chromatophores are small ovoid or discoid and are arranged in the peripheral layer of cytoplasm. They have chlorophyll a, chlorophyll e, carotenoids and lack pyrenoids.
10. Nuclei are found internal to chromatophores.
11. Large number of oil droplets are found which make the reserve food material of the thallus.
12. Reproduces asexually by multi-flagellate zoospores aplanospores and akinetes.
13. Sexual reproduction is oogamous. Sex organs are antheridia and oogonia.

1.9 ECTOCARPUS

Division — Phaeophyta
Class — Phaeophyceae
Sub-class — Isogeneratae
Order — Ectocarpales
Family — Ectocarpaceae
Genus — *Ectocarpus*

Identifying Characters

1. Thallus is branched, filamentous and heterotrichous.
2. Thallus bears unilocular and plurilocular sporangia.

ECTOCARPUS — Vegetative Structure

1. It is a marine brown alga found in cold seas of temperate and Polar Regions.
2. The thallus is filamentous, branched and heterotrichous showing the basal prostrate rhizoidal system and the aerial erect projecting photosynthetic system.
3. The prostrate system is irregularly and frequently branched. It is separate and attaches the thallus to the substratum, thus serves as holdfast.
4. Erect system arises from prostrate system and consists of many branched filaments, which are uniseriate i.e., single row of cells.
5. Branching is lateral and each branch tapers into a series of elongated cells forming a colourless hair like structure.
6. Cells are short and cylindrical. The cell wall is thick and consists of an outer layer of pectose and an inner layer of cellulose.
7. The cells are uninucleate. Chromatophores are golden coloured and are either disc shaped or band shaped. They contain chlorophyll a and b and the characteristic xanthophyll, fucoxanthin.

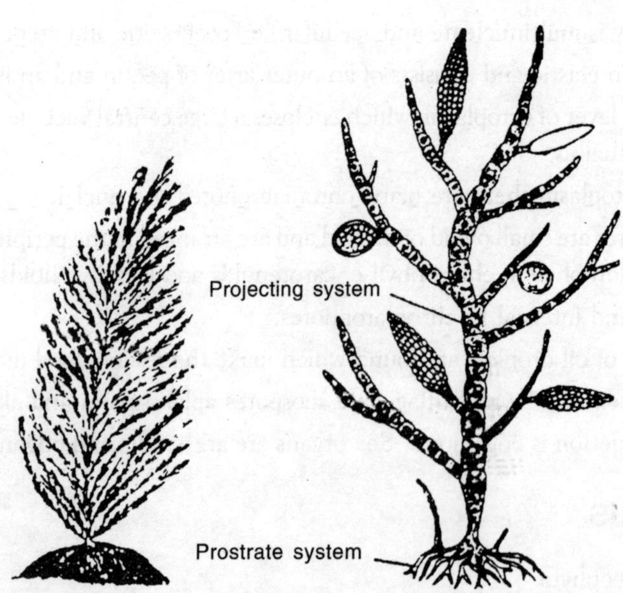

A. Showing habit **B.** Heterotrichous filament

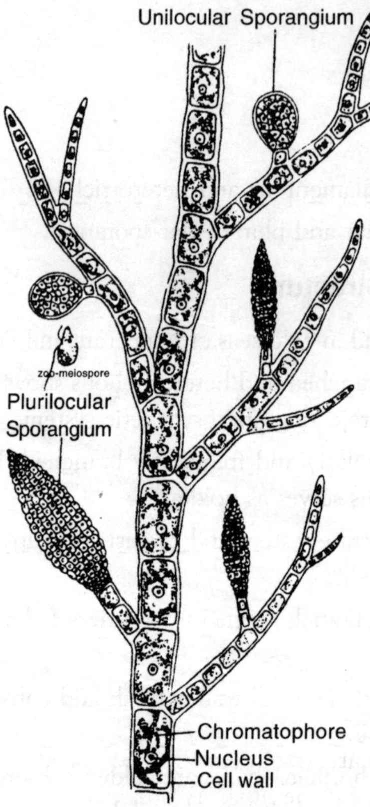

Fig. 1.17 Ectocarpus. Erect filament showing unilocular and plurilocular sporangia

8. Growth in the rhizodial protions is apical and upright portion is intercalary.
9. Genetically the thalli are of two kinds, haploid and diploid. Morphologically they are alike and alternate with each other in the life cycle.
10. The diploid thallus bears two kinds of sporangia, unilocular and plurilocular.
11. Unilocular sporangia after meiosis produces 32–64 haploid meiozoospores, which, give rise to haploid plants.
12. Plurilocular sporangia are elongated, cone like, and contain many cubical cells each of which produces diploid zoospores.
13. Haploid thalli produce biflagellate gametes in gametangia.
14. *Ectocarpus* shows isomorphic alternation of generation.

ECTOCARPUS—Plurilocular Sporangium

1. The given slide is the plurilocular sporangia of *Ectocarpus*.
2. The thallus of *Ectocarpus* is branched filamentous and heterotrichous.
3. Genetically these are two types of thalli, i.e., haploid and diploid.
4. Diploid thallus bears plurilocular sporangia, which are concerned with asexual reproduction.
5. Plurilocular sporangia are found at the end of small branchlets.
6. The terminal cell of the branchlet functions as the sporangial mother cell.
7. Sporangia are elongated cone like structures and are made up of many small cubical cells arranged in 20–40 vertical rows.
8. The cells of the sporangium are diploid and each of which produces a diploid zoospore, which on germination gives rise to diploid plant.
9. The zoospores are pear shaped and have two laterally placed flagella, one is pantonematic and the other acronematic.

ECTOCARPUS—Unilocular Sporangium

1. The given slide is the unilocular sporangia of *Ectocarpus*.
2. The thallus of *Ectocarpus* is branched filamentous and heterotrichous.
3. Genetically there are two types of thalli, i.e., haploid and diploid.
4. The diploid thallus bears unilocular sporangia.
5. These are borne singly on lateral branches.
6. Each sporangium is stalked globular or pear shaped structure having dense cytoplasm with many chromatophores and diploid nucleus.
7. The diploid nucleus divides meiotically into 32–64 haploid nuclei. Each uninucleate protoplast develops into a haploid meiozoospore.
8. Each meiozoospore is uninucleate, small pear shaped haploid structure with two laterally placed flagella, one is pantonematic and the other acronematic.
9. The meiozoospores germinate and develop into haploid sexual plants.

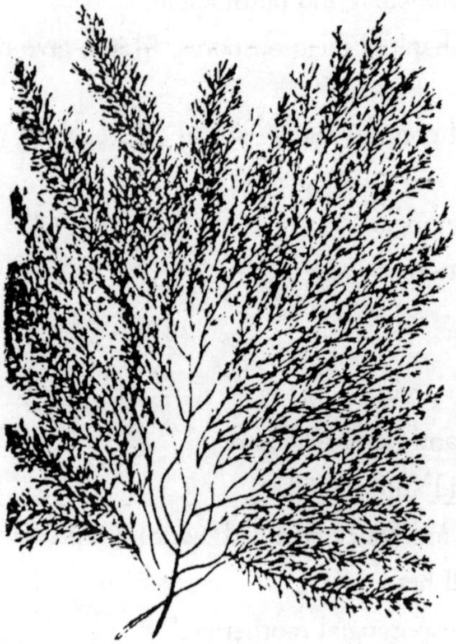

Fig. 1.18 Polysiphonia sp. showing habit

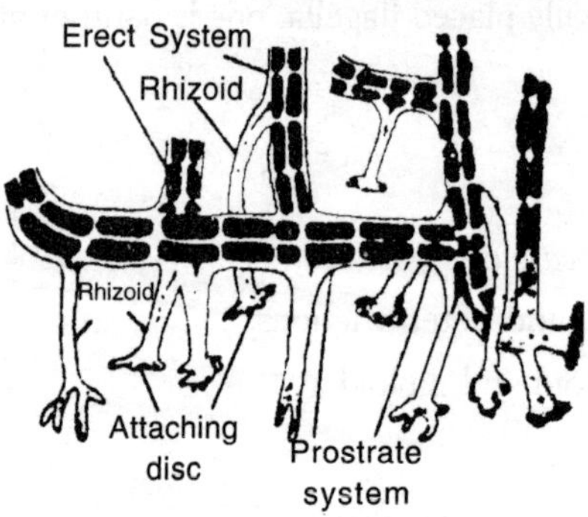

Fig. 1.19 Polysiphonia. Thallus showing erect and prostrate filament

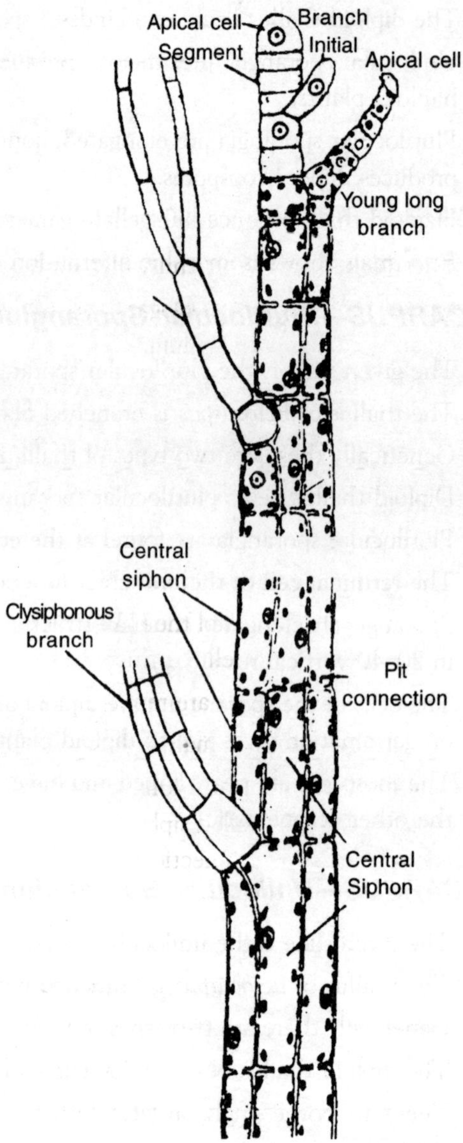

Fig. 1.20 Polysiphonia. A portion of thallus enlarged showing polysiphonous condition

1.10 POLYSIPHONIA

Division — Rhodophyta
Class — Rhodophyceae
Sub-class — Florideae
Order — Ceramiales
Family - Rhodomelaceae
Genus - *Polysiphonia*

Identifying Characters

1. The thallus is heterotrichus, filamentous and polysiphonous.
2. Sexual reproduction is Ooganeous. Male sex organ is called spermatangium and female sex organ is called carpogonium.
3. Haploid tetraspores are produced on tetrasporophyte.

POLYSIPHONIA—Vegetative Structure

1. It is a marine red alga and occurs in sublittoral zones of brackish estuaries and tidal pools.
2. Thallus is brownish red to dark purple red in colour.
3. Thallus is filamentous branched and heterotrichous.
4. The prostrate portion of the thallus creeps on the substratum and is anchored to the substratum by thick walled unicellular rhizoids, the tips of which are flattened and disc like.
5. The upright filaments are repeatedly branched and have delicate feathery appearance.
6. Upright system has a main axis having long branches and short branches or trichoblasts.
7. The main filament and long branches are polysiphonous, i.e., they consists of a system of parallel filaments called siphons, made up of elongated cells arranged one upon the other and connected by pit connections.
8. There is one axial siphon called central siphon, surround by 4–20 pericentral siphons.
9. The short branches (trichoblasts) are monosiphonous and dichotomously branched and bear sex organs.
10. Each cell is uninucleate with many discoid chromatophores arranged in the periphery of the cytoplasm. Cells are connected with each other by cytoplasmic pit connections.
11. Polysiphonia shows diplobiontic life cycle and is triphasic. It shows two diploid and one haploid individual. The individuals are:
    (a) Gametophyte—Free living haploid male and female plants having haploid sex organs spermatangia and carpogonia.
    (b) Carposporophyte—Diploid and dependent on gametophyte and produces diploid carpospores.
    (c) Tetrasporophyte—Diploid free living and produces haploid tetraspores.

POLYSIPHONIA—Spermatangia

1. The given slide shows the spermatangia, which are the male sex organs of *Polysiphonia*.
2. The thallus of *Polysiphonia* is heterotrichous, filamentous, polysiphonous and branched.
3. Male thallus bears spermatangia, which are closely packed and form a compact cone shaped structure on the short monosiphonous branches called male trichoblasts.
4. The male trichoblast consists of two basal cells constituting the stalk, which forks into two branches of which one is fertile and the other is sterile.
5. Spermatangia are spherical or oblong unicellular structures.
6. It has a large nucleus and colourless cytoplasm.
7. The spermatangial wall is thick and differentiated into 3 layers.
8. The uninucleate protoplasm of spermatangium produces a single male gamete called spermatium.
9. Spermatium is unicellular, spherical,non-motile structure and are liberated through a slit in the spermatangial wall.

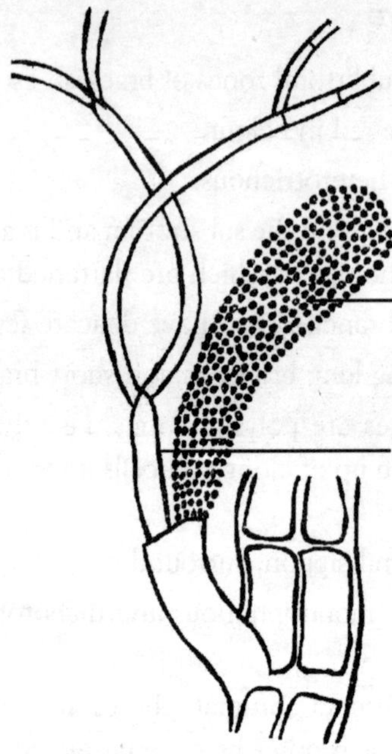

Fig. 1.21 Spermatangium

POLYSIPHONIA—Tetrasporangia

1. The given slide is *Polysiphonia* showing tetrasporangia.
2. The thallus of *Polysiphonia* is heterotrichous, filamentous, polysiphonous and branched.
3. It shows diplobiontic life cycle and is triphasic, i.e., three individuals in the life cycle.

4. Tetrasprophyte is the third individual in the life cycle. It is diploid and free living and bears tetrasporangia.
5. Tetrasporangia are found in longitudinal series and develop from the pericentral cells.
6. Tetrasporangia are small spherical bodies borne on short one celled stalk and are externally covered by two cover cells.
7. Each tetrasporangium possesses four tetrahedrally arranged uninucleate haploid tetraspores.
8. The tetraspores germinate to give rise to gametophyte.

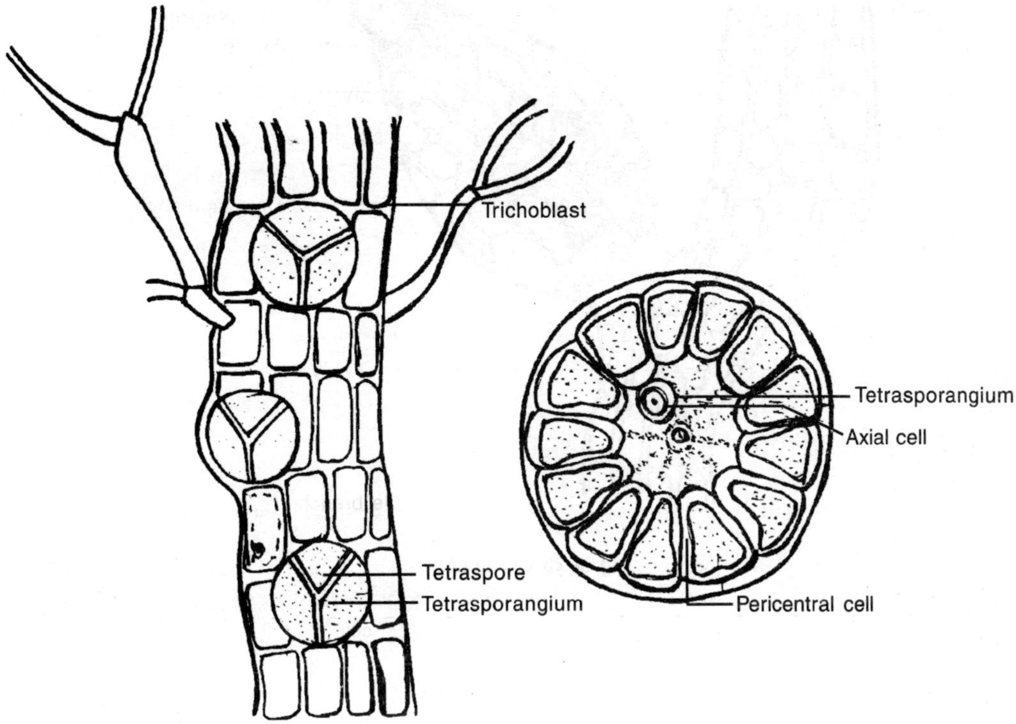

Fig. 1.22 Tetrasporangia

POLYSIPHONIA—Cystocarp

1. The given slide is the cystocarp or carposporophyte of *Polysiphonia*.
2. The thallus of *Polysiphonia* is heterotrichous, filamentous polysiphonous and branched.
3. *Polysiphonia* shows diplobiontic life cycle and is triphasic, i.e., three individuals in the life cycle.
4. Cystocarp is the second individual in the life cycle. It is partly diploid and is dependent on the gametophyte.
5. It is an oval or urn shaped structure attached to the gametophytic filament (haploid). It opens to the exterior by a pore called ostiole.

6. Wall of the cystocarp is called pericarp and is made up of single layer of cells, which are haploid.
7. Cystocarp consists of a placental or fusion cell at the base.
8. From the placental cell many diploid filaments called gonimoblast filaments arise, the terminal cells of which bear carposporangia, each of which bears a single diploid carpospore.
9. Carpospore on germination produces diploid tetrasporophyte.

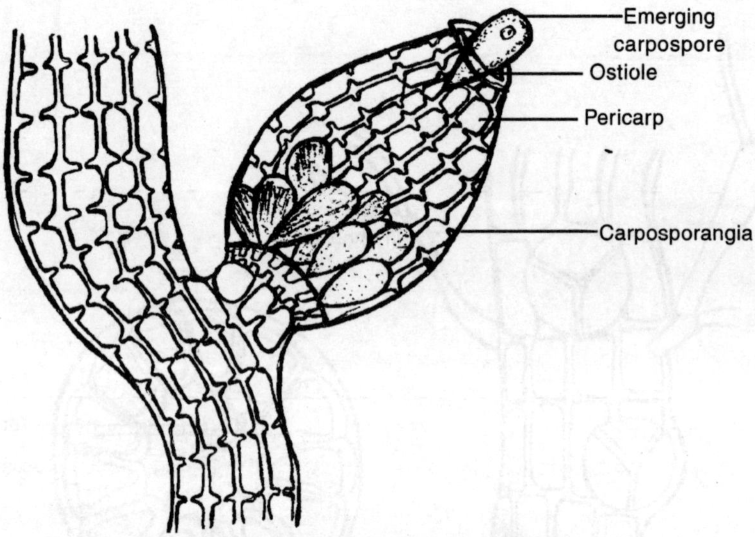

Fig. 1.23 Cystocarp

2 CHAPTER FUNGI

2.1 INTRODUCTION

The word 'fungus' (pl., fungi) is derived from Latin which means mushroom. The Greek word Mycology (Mykes = mushroom; logos = study or discourse) relates to the study of mushrooms.

Alexopolus (1982) in his book "Introductory Mycology" defined fungi as "Eukaryotic, spore bearing achlorophyllous organisms that generally reproduce sexually and asexually and whose filamentous branched somatic structures are typically surrounded by cell walls containing chitia or cellulose or both of these substances, together with many other complex organic molecules".

The fungi grow in diverse habitats. The are found in almost every available habitat on earth where organic material is present. They are universal in their distribution. The plant body of the fungi is mostly an undifferentiated thallus. It may be unicellular or filamentous thallus known as mycelium. The individual filaments in the mycelium are called hyphae. The hyphae are branched, narrow colourless or coloured rigid tube like structures. The hyphae may be septate or aseptate. The septate hyphae may be uninucleate or binucleate. The cell wall is made up of chitin except in Oomycetes.

Fungi reproduce both by asexual and sexual methods. Asexual reproduction takes place by the production of spores like aplanospores, zoospores, chlamydo spores, conidia etc.

Sexual reproduction involves the union of two nuclei, gametes or sex organs. It includes three phases: plasmogamy, karyogamy and meiosis. Asexual reproduction is effected by different methods like planogametic copulation, gametangial contact, gametangial copulation Cystopus, spermatisation and somatogamy.

2.2 LEAF OF BRASSICA (MUSTARD) – Infected by Cystopus

Division — Eumycota

Sub-division — Mastigomycotina

Class — Oomycetes

Order – Peronosporales

Family — Albuginaceae
Genus — *Albugo* (=*Cystopus*)
Species — *Albugo candida*

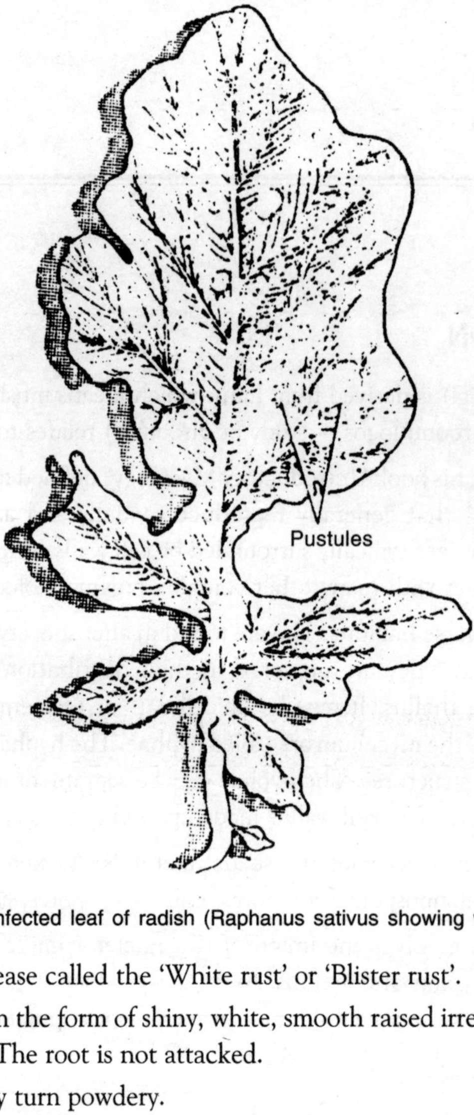

Fig. 2.1 Cystopus. Infected leaf of radish (Raphanus sativus showing white pustules)

1. *Cystopus* causes a disease called the 'White rust' or 'Blister rust'.
2. The disease appears in the form of shiny, white, smooth raised irregular blisters or patches on the leaves or stems. The root is not attacked.
3. The patches gradually turn powdery.
4. The epidermis is ruptured by the pressure of sporangia, which on coming out give the appearance of white powdery mass.
5. The leaves in some cases become fleshy and thickened, inrolled and in severe infection, the entire plant may be dwarfed.
6. The flowers and stems show hypertrophy.
7. The distorted and hypertrophy inflorescence are called stag heads.

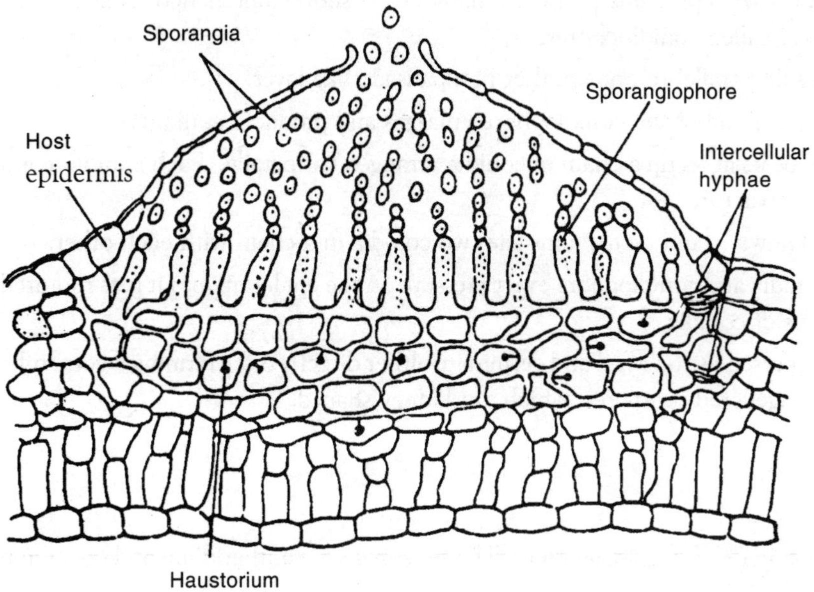

Fig. 2.2 *Albugo candida*. A section of host leaf passing through the sporangial sorus

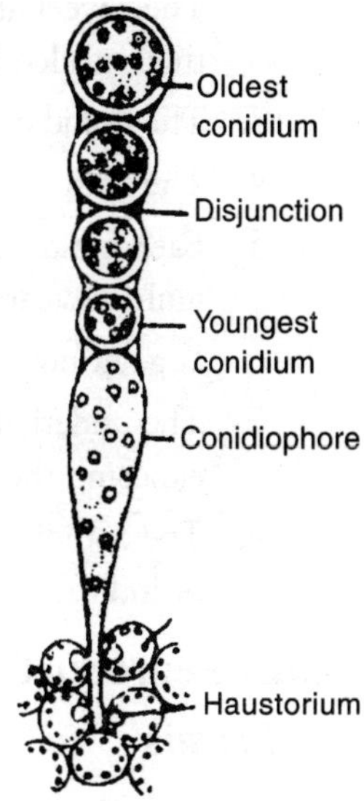

Fig. 2.3 *Cystopus*. A single conidiophore with conidia

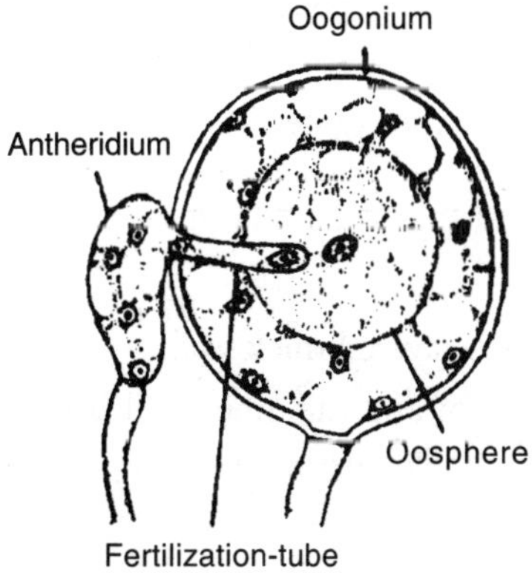

Fig. 2.4 *Cystopus*. Sex organs

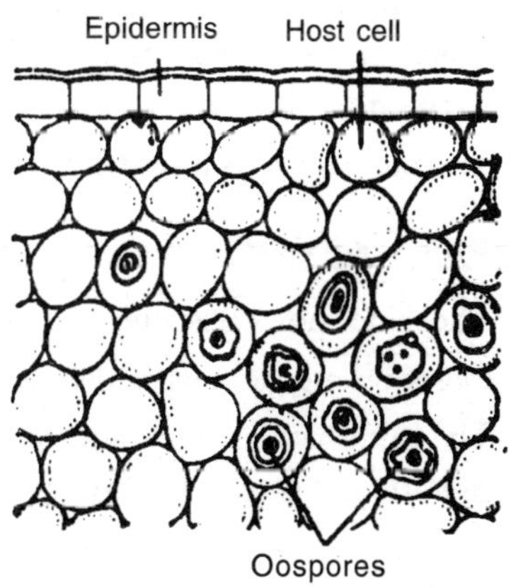

Fig. 2.5 *Cystopus*. T.S. host stem showing oospores

CYSTOPUS — Asexual Reprodution

1. It takes place by means of conidia.
2. The mycelium below the epidermis produces many erect, short unbranched, club shaped, thick walled hyphae called conidiophores.
3. The conidiophores lie parallel to one another in a palisade like layer.
4. The thin-walled apical end of the conidiophore enlarges and produces conidium.
5. Each condiophore bears at its tip a chain of conidia arranged basipetally. Each conidium is a multinucleate oval structure.
6. A gelatinous pad known as disjunctor joins the two conidia in a chain with each other.
7. The underlying conidia and conidiophore exert pressure on the epidermis and it gets ruptured exposing the chains of conidia.
8. The conidia are dispersed by the wind and germinate either directly or by forming a germ tube or indirectly by biflagellated zoospores, which are kidney shaped.

Sexual Reproduction

Sex Organs

1. Sexual reproduction in *Cystopus* is oogamous. The sex organs are antheridium and oogonium developed on different thalli.
2. The sex organs develop side by side and such arrangement is called paragynous.
3. The oogonium is globular, multinuclear and contains a large amount of food material and at the base it bears a septum.
4. The oogonial cytoplasm is differentiated into 2 regions outer vacuolated periplasm and central dense ooplasm.
5. The ooplasm has a single functional nucleus and reserve food material.
6. The antheridium is an elongated club shaped cell. It produces a slender fertilization tube which grows through the oogonial wall and the periplasm and penetrates deeply into the ooplasm or oosphere.
7. The male nucleus fuses with the female nucleus and forms the zygote.

Oospores of Cystopus

1. The slide shows oospores of *Cystopus*.
2. Oospores are globular bodies and are surrounded by a two-layered thick wall.
3. The outer wall is thick and warty and is called exosporium and the inner wall is thick and smooth and is called endosporium. Sometimes a thick middle layer is present.
4. The diploid nucleus undergoes meiosis and after repeated mitotic divisions produces 32 nuclei.
5. In this 32 nucleate state it enters resting stage and tides over unfavourable period.

2.3 MUCOR

Asexual Reproduction

Division — Eumycota

Sub-division — Zygomycotina

Class — Zygomycetes

Order — Mucorales

Family — Mucoraceae

Genus — *Mucor*

1. *Mucor* is a saprophytic fungus growing on food stuff that are rich in starch and is called as black mould.
2. The plant body is a mycelium, which is aseptate, coenocytic and well branched.
3. It reproduces asexually by non-motile spores, which are produced in small, round and black bodies called sporangia.
4. The sporangia are produced terminally and singly on the sporangiophore.
5. Each sporangium develops at the tip of long, erect, unbranched, hypae growing vertically from the prostrate mycelium.
6. The young sporangium contains a number of nuclei, cytoplasm and reserve food material. The cytoplasm cleaves into multinucleate non-motile spores.
7. At maturity the spores are embedded in the liquid secreted by the columella, which develops into bright glistening sporangial drop. Agents like insects disperse the spores.
8. Under favourable conditions the dispersed spores germinate to form new mycelium.

Sexual Reproduction

1. Species of mucor are dioecious and exhibit heterothallism.The method of plasmogamy is gametangial copulation and the two fusing gametangia (male and female) belong to two different strains (+ and –) and are morphologically similar but physiologolically different .
2. The two compatible hyphae, which are called zygophores are attracted to each other.
3. The zygophores meet and put out a copulating branch at the point of contact called progametangia.
4. It develops a terminal cell called gametangium and the proximal part called suspensor.
5. When the gametangium matures the intervening walls between them dissolve and the nuclei fuse in pairs.
6. The combined protoplast in the fusion cell containing a number of diploid nuclei is called zygospore.
7. The zygyospores secrete a thick two layered wall around it, the outer layer is dark and warty called exine and the inner layer called intine.
8. After a period of rest, zygospore divides mitotically and puts out a germ tube to form the zygosporangium.
9. The sporangium contains haploid spores and each spore after liberation germinates to form a new mycelium.

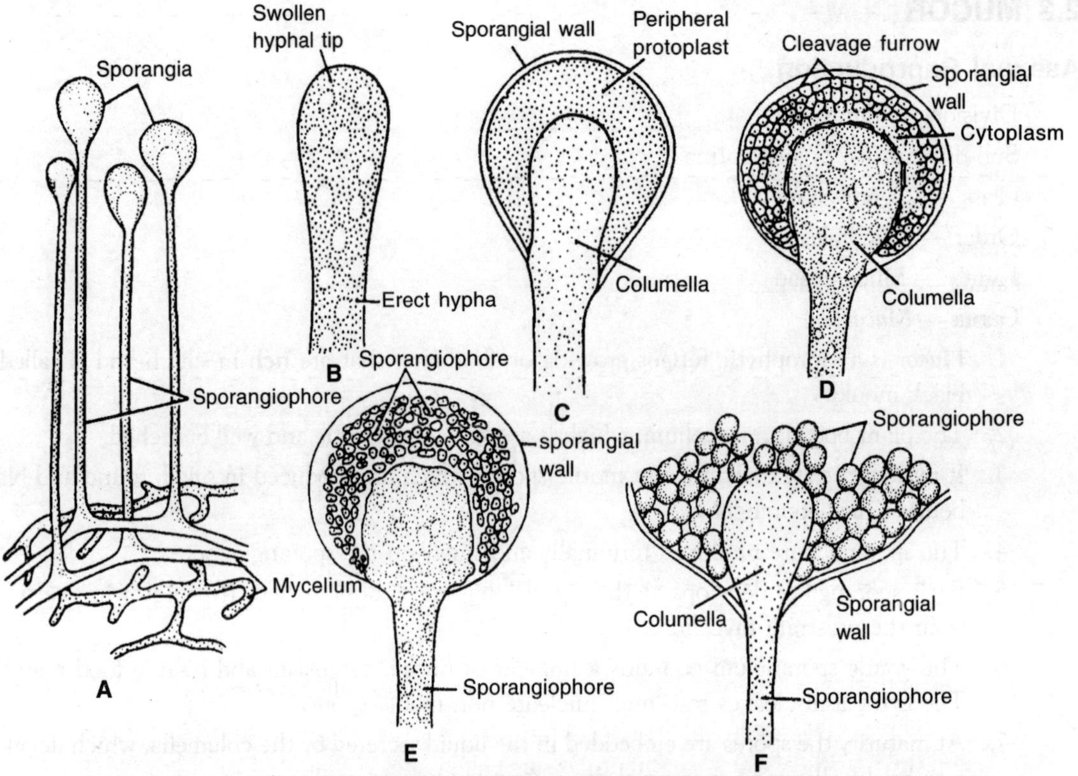

Fig. 2.6 *Mucor.* **A.** Hyphae bearing sporangia, **B-D.** Stages in the development of sporangium, **E.** Mature sporangium, **F.** Dehisced sporangium

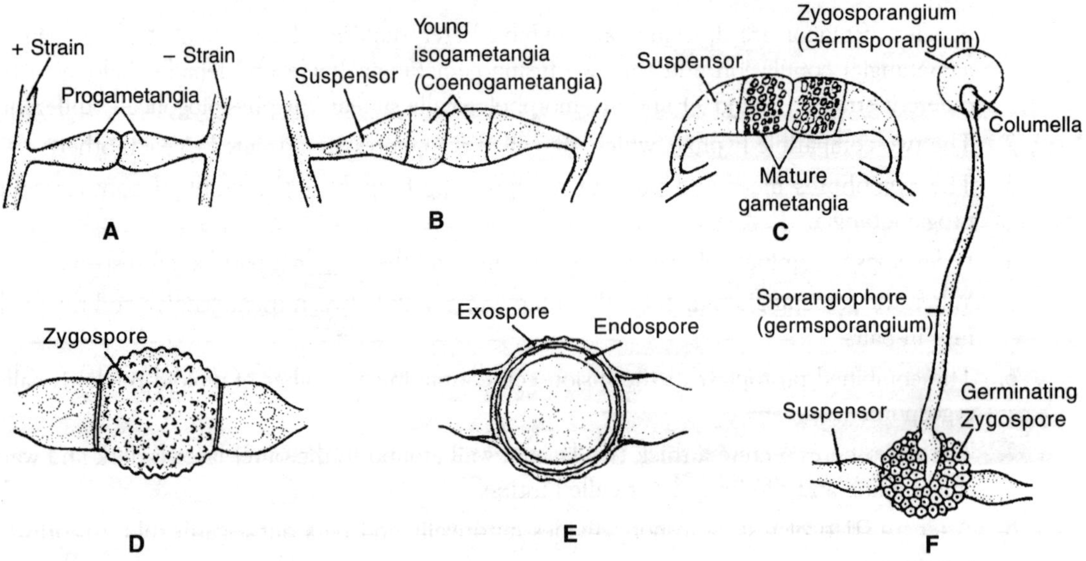

Fig. 2.7 *Mucor.* Sexual reproduction; **A-C.** Stages in the development of gametangia, **D-E.** Zygospore, **F.** Zygosporangium

2.4 PENICILLIUM

Asexual Reproduction

Division — Eumycota
Sub-division — Ascomycotina
Class — Plectomycetes
Order — Eurotiales
Family — Aspergillaceae
Genus — *Penicillium*

1. They are commonly called green or blue molds.
2. The mycelium is branched, septate and each cell is multinucleate. It is a saprophyte on decaying fruits and vegetables.
3. Asexual reproduction takes place by non-motile spores called conidia borne on conidiophores exogenously.
4. Conidiophore arises as an erect tubular hyphal outgrowth from any cell of mycelium.
5. The condiophore branches once or twice and it is termed primary, secondary or tertiary branch.
6. The condiophore bears a tuft of flask shaped structures called sterigmata.
7. In the branched conidiophores, the branches, which bear the sterigma, are called metulae, and the branches, which bear metulae, are called rami.
8. The conidia are formed within the tip of flask shaped sterigma by swelling.
9. The conidia are borne in basipetal chains.
10. Each conidium is ovoid or globular with smooth or rough wall, which is uninucleate and bluish or greenish in colour.
11. The conidiophore with the branches is in the form of artist's brush and that is why known as *Penicillium*. (Generic name derived from it)
12. Conidia are dispersed by wind and on falling on a suitable substratum germinate into new mycelia.

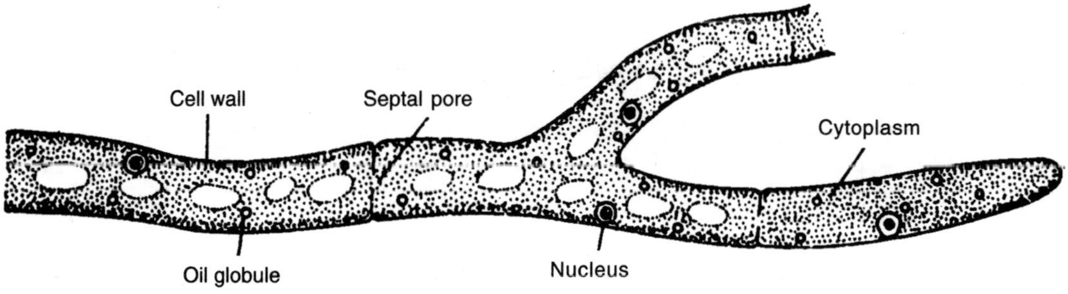

Fig. 2.8 *Penicillium.* A portion of a hypha enlarged to show structure

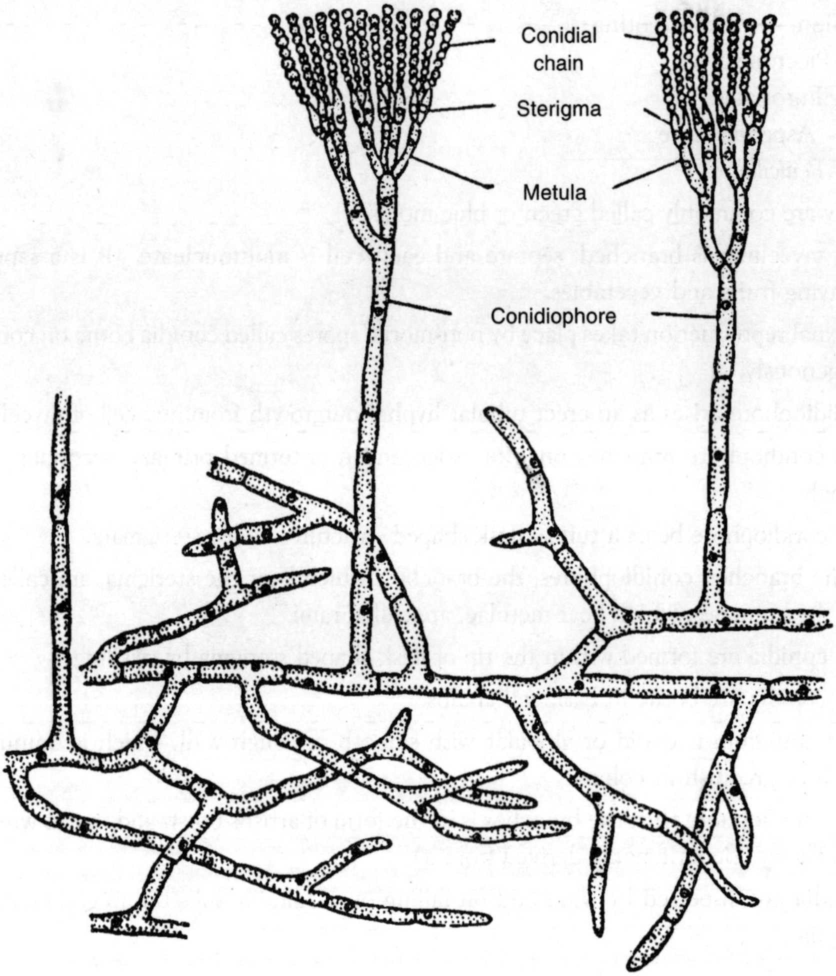

Fig. 2.9 *Penicillium.* Asexual reproduction

2.5 PEZIZA—APOTHECIA

Division — Eumycota
Sub-division — Ascomycotina
Class — Discomycetes
Order — Pezizales
Family — Pezizaceae
Genus — *Peziza*

1. *Peziza* is a saprophytic fungus occurring on animal dung, decaying wood or heavily manured soil.

Fig. 2.10 Peziza apothecia

2. Apothecia are the fleshy fruiting bodies (fruitfication) of *Peziza*.
3. The apothecia are aerial, short lived, sessile or short cup shaped structures. They vary in size from 2 cm to several inches.
4. It varies in colour from pale brown to black to red, orange, yellow or colourless.
5. The fungus has well-developed, perennial network of hyphae that are branched, septate, and uninucleate. The mycelium is hidden and ramify in the substratum and extract nourishment from it.
6. Fruiting bodies are produced above the ground at the time of sexual reproduction.
7. Apothecium chiefly consists of three parts namely hymenium, hypothecium and excipulum.
8. Hymenium is the fertile layer that lines the cavity of the cup or surface of the disc and is called the thecium, which is dark brown in colour.
9. Hypothecium is below the hymenium and consists of network of hyphae, runs horizontally.
10. Excipulum is the haploid part of ascocarp.

PEZIZA – V.S. of Apothecium

1. The ascocarps are of apothecium type where the hymenium is exposed.
2. The apotheium is disc-shaped or cup-shaped and either stalked or sessile.
3. The apothecium is made up of three parts viz., hymenium, hypothecium and excipulum.
4. Hymenium is the fertile layer, which lines the cavity of the cup. It ęncloses the sac-like structures the asci which are parallel and in palisade like layer.
5. The asci enclose 8 ascospores within the cytoplasm.
6. Intermingled with the asci are slender, septate, sterile hyphae called paraphysis, which protect and support the asci.
7. In some cases the tips of paraphysis extend beyond the asci and form a lining layer called epithecium.
8. Beneath the hymenium is the hypothecium from which the paraphysis develop. It is a horizontally running network of hyphae.
9. Beneath hypothecium is the fleshy part of ascocarp, which forms the base and sides and is called excipulum. It is a pseudoparenchymatous tissue, which is further differentiated into outer ectalexcipulum and inner entalexcipulum.

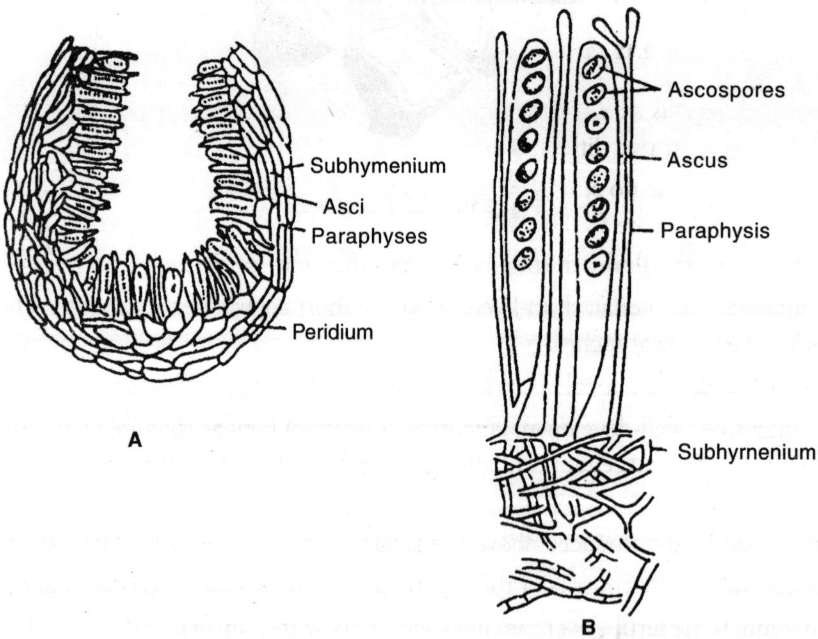

Fig. 2.11 Peziza. **A.** V.S. entire fructification (apothecium); **B.** Portion of the same to show details (Digrammatic)

2.6 RED RUST ON WHEAT PLANT (Primary Host)

Division — Eumycota
Sub-division — Basidiomycotina
Class — Teliomycetes
Order — Uredinales
Family — Pucciniaceae
Genus — *Puccinia graminis*

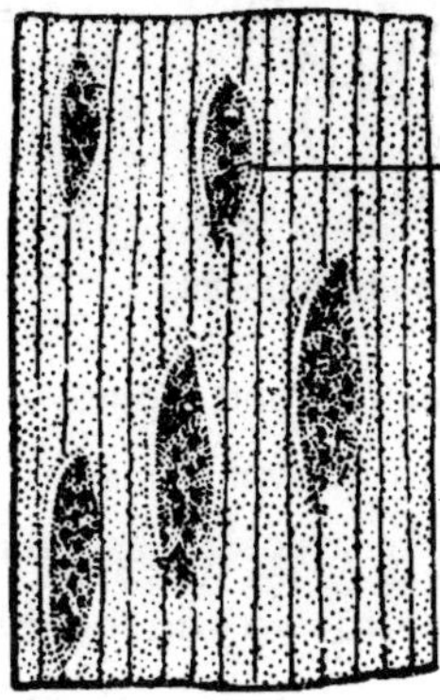

Fig. 2.12 *Puccinia.* A portion of leaf showing uredo pustules

1. *Puccinia graminis* is an obligate parasite. It is a heteroecious species because they complete their life cycle on two different host plants.
2. The uredo and teleuto stages are found on the primary host, the wheat plant and pycnidial and aecidial stages are found on Barberry plant.
3. *Puccinia graminis* infects the wheat plant and causes the Black stem rust disease. The symptoms of the disease are seen on the leaves, leaf sheaths and stem. Brown to black or linear lesions are produced on leaves, leaf sheaths and stems which in severe infection collapse to form large patches.
4. The infection starts by the germination of the aecidiospore on the stem. The germ tube penetrates the host cell wall and enters the tissue through the stoma. The germ tube elongates and develops into the secondary or dikaryotic mycelium. After a few days of its infection the mycelium aggregates and forms the uredosori and externally they are seen as brown streaks.

PUCCINIA GRAMINIS

T.S. of leaf through Uredia or Uredosorus

1. *Puccinia graminis* is a heteroecious parasite, which needs two host plants, wheat the primary host and Barberry the secondary host.
2. The dikaryotic mycelium grows on wheat plant and haploid mycelium on Barberry plant.

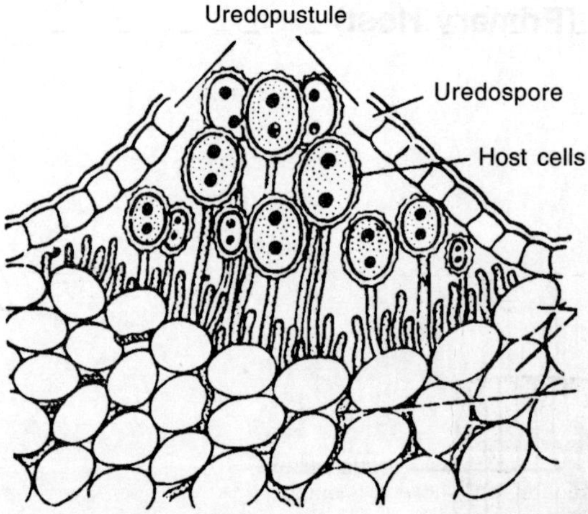

Fig. 2.13 *Puccinia.* T.S. of wheat leaf through uredo-pustule

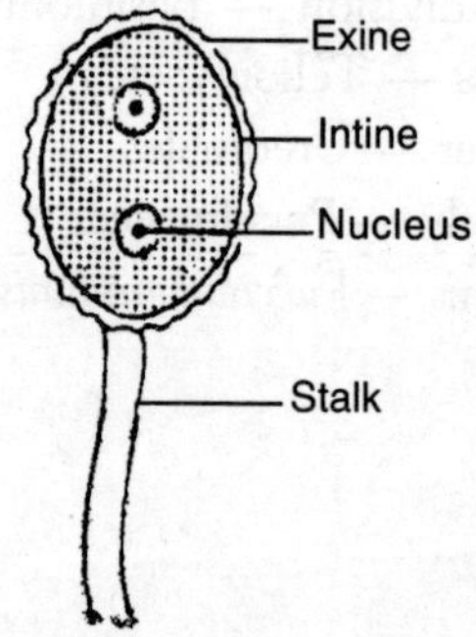

Fig. 2.14 *Puccinia.* Uredospore

3. The mycelium is intercellular, branched, and septate and each cell is dikaryotic.
4. The dikaryotic mycelium aggregates as hyphal masses beneath the epidermis of stem, leaf, sheaths or glumes and forms mat like structures called uredia.
5. From the uredia vertical, stalk-like structures arranged in the palisade manner are produced.
6. The tip of each hyphae swells to form a uredospore and the uredospores in groups are called uredosori.
7. Each uredospore is stalked, unicellular, oval, binucleate structure with thick wall. They are rust red in colour. The outermost layer is echinulate with four equatorially arranged thin areas on it called the germ pores.
8. The uredospores exert pressure on the overlying epidermis and ruptures in the form of slits or blisters.
9. The uredospores are carried away by wind to the healthy plant and germinate on it causing infection again within few hours.

TELIA OR TELEUTOSORI

T.S. of Wheat Leaf Through Telia or Teleutosori

1. Among the uredosori, some teleutospores begin to develop at the end of growing season. The teleutosori or teleuto pustules appear on leaves, leaf sheaths and stem as black oval pustules that fuse to form patches in severe infection.
2. As the season progresses more and more teleutospores are produced and finally the entire sori contain only teleutospores and the sori is called teleutosori.
3. Each teleutospore is black or brown coloured, stalked, two-celled, spindleshaped structure with a constriction at the septum. The wall is thick and smooth; there is a single germ pore in the wall of each cell. These are resting, resistant spores enabling the fungus to tide over unfavourable season.

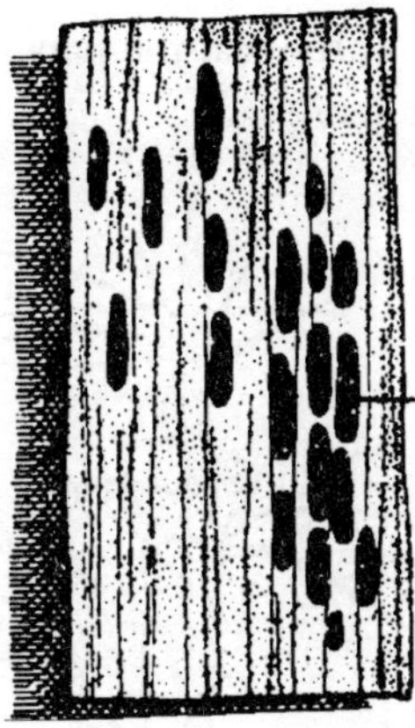

Fig. 2.15 *Puccinia.* A portion of leaf showing teleutopustule

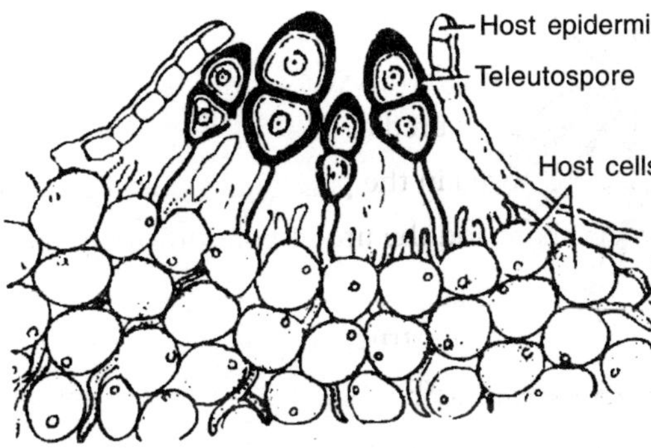

Fig. 2.16 *Puccinia.* T.S. of wheat leaf through teleuto-pustule

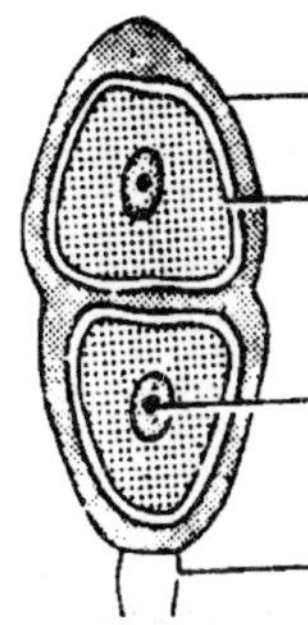

Fig. 2.17 *Puccinia.* Teleutospore

4. The teleutospores exert pressure on the overlying epidermis and ruptures it and the spores are exposed.
5. The two cells are binucleate at first but later on the two haploid nuclei fuse together to form diploid cells.
6. The teleutospores with a diploid nucleus in each cell represents the reduced diplophase.
7. The teleutospore undergoes resting period in this condition attached to the wheat straw in the soil.
8. With the return of favourable conditions the teleutospores germinate and give rise to the basidial stage. The diploid nucleus undergoes meiosis and gives rise to four haploid basidiospores which are borne on the sterigma of the epibasidium.

PYCNIDIAL OR SPERMAGONIAL STAGE ON BARBERRY PLANT

1. The basidiospores germinate on the leaf of secondary host i.e., Barberry plant and form the haploid mycelium.

2. After four days of infection on the leaf the haploid mycelium begins to collect beneath the upper epidermis and forms a dense mat.

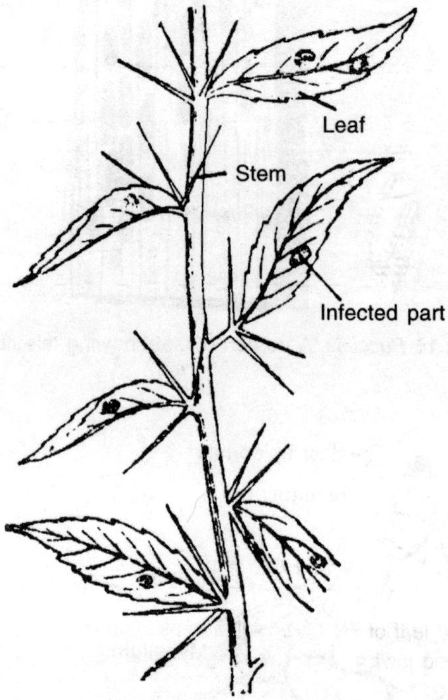

Fig. 2.18 *P.* graminis on Berberis plant.

3. The dense mat cuts into flask-shaped structures called spermagonia or pycnidia and they are either of + strain or - strain.
4. Each spermagonium consists of a wall surrounding a cavity with an opening called ostiole.
5. Three different types of hyphae arise from the wall of the spermagonium. They are spermatiophores, periphysis and receptive hyphae or flexous hyphae.
6. Spermatiophores are fine, short, uninucleate hyphae which are packed closely in a palisade-like layer.
7. Each spermatiophore abstricts from its tip a number of spermatia which are non-motile.
8. The periphysis are long, delicate, sterile hyphae which develop near the ostiole from the spermagonial wall.
9. Among the spermatophores another kind of hyphae which are slender, delicate, cylindrical with blunt fine tips develop and they are called receptive hyphae or flexous hyphae.
10. The receptive hyphae are spermatised by spermatia of opposite strain with the help of insects.

PUCCINIA—AECIDIAL CUP ON BARBERRY PLANT

1. The aecidia are generally present on the lower surface of leaf and thus both pycnidia and aecidia are seen in the same section of the host leaf.

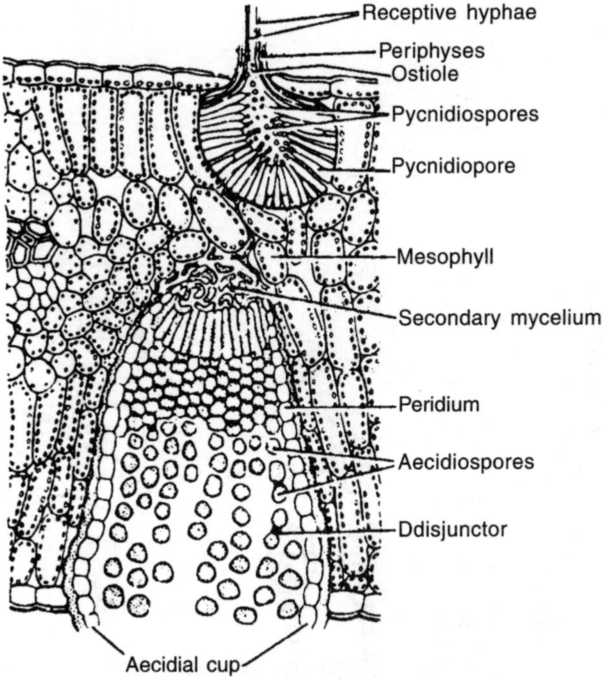

Fig. 2.19 *Puccinia.* T.S. of infected leaf of *Berberis* showing pycnidial and aecidial cups on upper and lower epidermis respectively

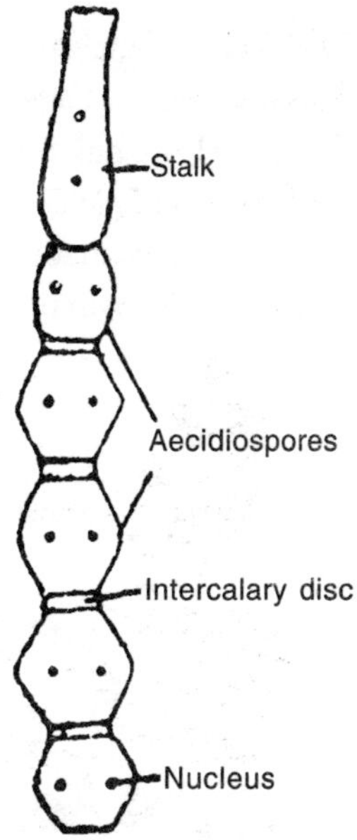

Fig. 2.20 Many aecidiospores on stalk

2. The aecidial cup is formed by the dikaryotic mycelium.
3. The dikaryotic mycelium accumulates and forms masses of uninucleate cells and they are called aecidial primordia.
4. The aecidial primordia lies in the substomatal chamber.
5. The cells in the basal layer of the aecidial primordia become binucleate and are called aecidiospore mother cells.
6. The aecidiospore mother cell increases in length and the nucleus divides conjugately.
7. A small daughter cell is cut off which is called terminal cell. The terminal cell further undergoes another conjugate division and gives rise to bigger aecidiospore and smaller intercalary cell or disjunctor.
8. Each aecidiospore undergoes a series of such division and gives rise to chains of aecidiospores.
9. The marginal cells of the aecidium divide repeatedly and form a protective wall which covers the spores called peridium.
10. Each aecidiospore is polyhedral, thin walled, unicellular binucleate and orange coloured.
11. The aecidiospores are dispersed by air currents and germinate on the Wheat plant and develops into dikaryotic mycelium.

2.7 ALTERNARIA

Division — Eumycota
Sub-division — Deuteromycotina
Class — Hyphomycetes
Order — Moniliales
Family — Dematiaceae
Genus — *Alternaria*

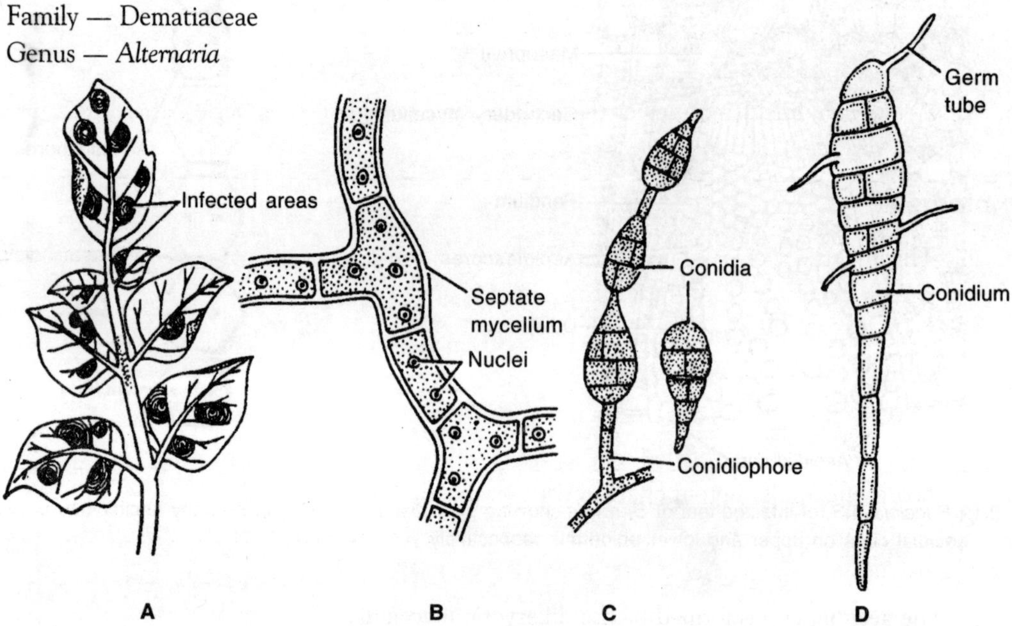

Fig. 2.21 *Alternaria.* **A.** Infected leaves of potato, **B.** Mycelium, **C.** Condiophore with conidia, **D.** Germination of conidium

Vegetative Structure

Cut a section of host, stain the section in cotton blue and mount in lacto phenol for study of mycelium.

1. Mycelium is intercellular or sometimes intracellular and yellowish brown in colour.
2. Hyphae are multicellular and branched.
3. Hyphae are also septate. Each cell is uninucleate.

Conidia

1. The conidia also are usually yellowish brown in colour.
2. The conidia may either be single or in chains. These are borne on conidiophores, which are not much distinguishable from the vegetative hyphae emerging through stomata.
3. They are long, dark coloured, muciform (beaked), multicellular and dictyosporous i.e., spindle shaped or ovoid with both transverse and longitudinal septa.
4. The perfect stage of this form genus whenever known bongs to loculoascomycetes—genus *Pleospora*.

Host Diseases

Majority of the species occur as weak parasites but some are saprophytes.

1. Of all the species, A. *solani* is economically very important, as it attacks potato plant (*Solanum tuberosum*) and causes early blight of potato. In Indian plains, early blight is more destructive than late blight.
2. A.*brassica and* A. *brassicola* cause gray and dark leaf spots of *Brassica* sps.
3. A.*triticina* causes leaf blight of wheat (*Triticum sps.*)
4. A. *alternata* infects sun flower (*Helianthus annus*) and large number of other host causing leaf spot disease.

Symptoms :

1. The disease appears on the leaf lets 3–4 weeks after the crop is sown as small isolated, scattered pale brown to dark spots, oval or angular in shape mostly up to 3–4 mm. in diameter.
2. Each spot is usually delimited by a narrow chlorotic marginal zone, which fades into the normal green. The chlorotic zone increases with increase in size of the spot.
3. The lower leaves are affected first and the disease progresses upwards.
4. The necrotic tissue of the spot shows series of concentric ridges, which produce a target-board effect, a symptom characteristic of this disease.
5. The number of spots on the leaf lets may be few but may increase in size and number if conditions are favorable.
6. In the advanced stage the number of spots is numerous and the leaf shows signs of old age and droops.
7. The spots become hard in dry weather and the leaflets curl.
8. During humid conditions the diseased areas coalesce and big rotting patches appear on the leaf surface.
9. In severe cases of infection the leaves dry up, shrivel and drop off.
10. Stems and petioles may also show brown to dark lesions, which may finally lead to collapse of the entire plant.

Disease Cycle

The source of primary inoculum is the infected plant debris such as the dried leaf, stems and potato tubers. Primary infection may be brought by conidia or the mycelium from the infected debris in the soil. The conidia germinate and the germ tubes enter into the host plant through stomata. The incubation period varies from 48–72 hours. The primary infection appears on the foliage as leaf spots within two or three days under favourable conditions.

Secondary infection produces from the primary spots though within 5–7 days after infection. Conidia develop when the spots are about 3mm in diameter. The mature conidia are dispersed through the agency of air currents, water and insects.

Control Measures

1. Crop rotation, field sanitation and burning of the plant debris.
2. Applications of Bordeaux mixture and other fungicides

2.8 LICHENS

Lichens are group of organisms, composite or dual in nature. The thallus is formed by the association of an alga and a fungus. The fungal partner is called a mycobiont and the algal partner as phycobiont.

Lichens are cosmopolitan in distribution and grow on decaying wood, bark of trees, rocks and in soils rich in organic matter.

CRUSTOSE LICHENS

The specimen shows the following features.

1. The crustose lichens are hard, granular crusts on rocks and bark of the trees.
2. These adhere very closely and firmly to the substratum.
3. The thalli are partially or completely embedded in the substratum.
4. These lichens are generally ash coloured, but sometimes may vary.
5. The surface of the thallus is generally divided into polygonal areas called areolae.
6. Internally the thallus is poorly differentiated.
7. The tissue forming the thallus is arranged in strata one above the other.
8. The upper cortex may be rudimentary or highly developed.
9. Algal layer lies just beneath the upper cortex, and shows closely associated algal filaments and fungal hyphae.
10. Beneath the algal layer lies the medulla composed of loose tissue of branched hyphae.
11. The lower cortex lies next to medulla and may be well developed or entirely absent.
12. Some of the common examples are *Lecidea, Graphis, Leconora.*

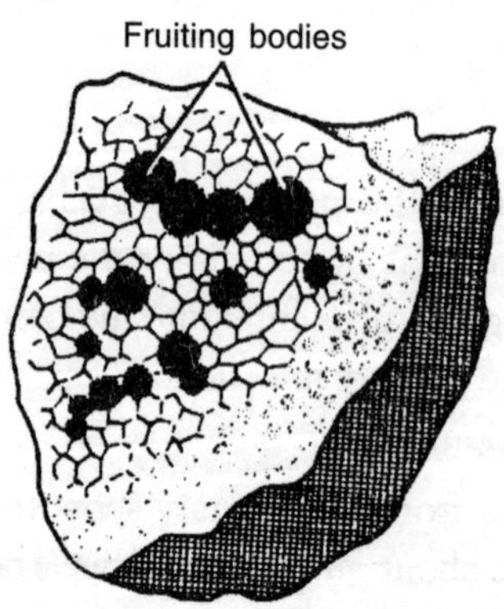

Fig. 2.22 *Lecidia platycarpa.* Crustose ascolichen growing on rocks : External features and fruiting bodies

FOLIOSE LICHENS

The given specimen shows the following features.

1. The thallus is a flat, leaf like, lobed or deeply incised and attached to the substratum at certain points by rhizine.
2. The rhizine are rhizoid-like out growths which develop from the lower surface.
3. The thallus is greyish or brownish in colour.
4. Some small, hard, dark and gall-like outgrowths called cephalodia may also be present.
5. Internally the thallus shows four distinct regions or zones. These are upper cortex, gonidial layer or algal layer, medulla and lower cortex.
6. Upper cortex may or may not be bounded by epidermis-like layer of hypha.
7. Breathing pores may also be present on the epidermis which help in gaseous exchange.
8. The upper cortex consists of vertical hyphae without intercellular spaces filled with gelatinous materials.
9. Beneath the upper cortex is algal layer consisting of numerous algal cells held together in the network of fungal hyphae.
10. Below the algal layer is the medulla made up of very loosely interwoven hyphae.
11. On the lower side is the lower cortex made of compact cells lying either parallel or perpendicular to the lower surface.
12. Cells of the lower cortex produce rhizine.

Examples: *Parmelia, Physcia, Collema, Leptogium.*

Fig. 2.23 *Parmelia flavicens.* Foliose ascolichen. A thallus showing external features and apothecia

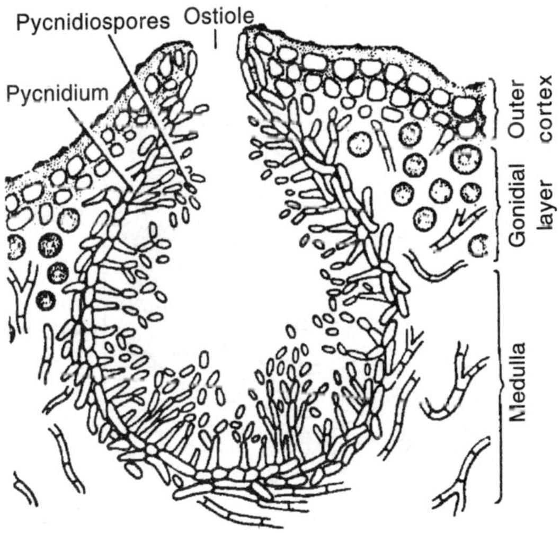

Fig. 2.24 *Physcia.* V.S. pycnidium to show pycnospores

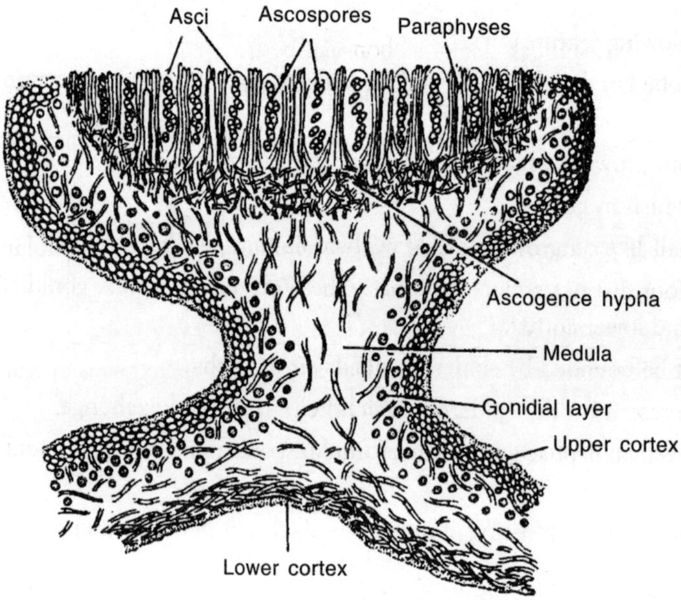

Fig. 2.25 *Physcia.* V.S. apothecium for the study of ascospores

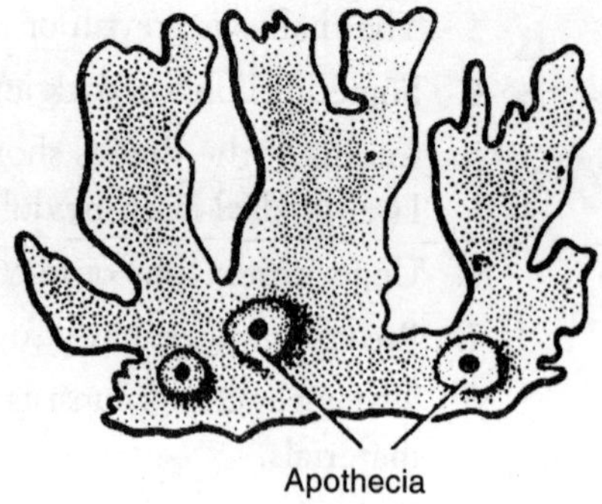

Fig. 2.26 *Physcia.* Thallus showing apothecia

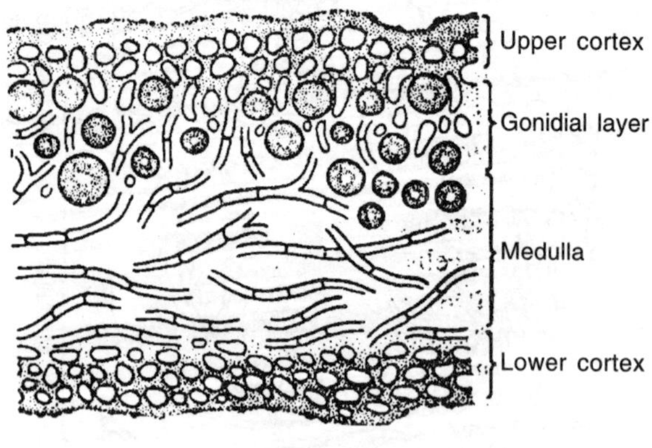

Fig. 2.27 *Physcia.* V.S. thallus to show heteromerous internal structure

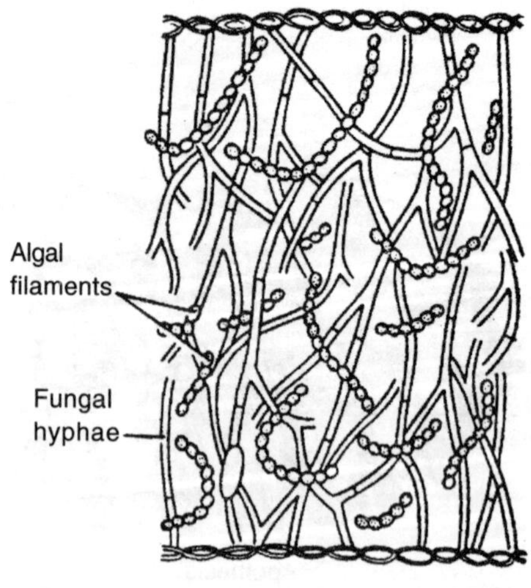

Fig. 2.28 *Collema.* Homoiomerous thallus showing internal structure

FRUITICOSE LICHENS

1. The thallus appears shrubby with cylindrical, flat or ribbon-like body.
2. It is upright, branched and pendulous and attached to the substratum by rhizoid-like structures forming a disc.
3. Internally the thallus shows four distinct regions, outer most cortex, the algal zone, medulla and centrally located chondroid axis.
4. The cortex is made up of closely packed interwoven fungal hyphae and the intercellular spaces may be absent.
5. The algal zone consists of cells of unicellular green alga.
6. Medulla is present beneath the algal zone and consists of algal cells loosely mixed with fungal hyphae.
7. The central choindroid axis is made up of longitudinally arranged, compact, thick walled and closely grouped fungal hyphae.

Examples: *Usnea, Cladonia, Alectonia.*

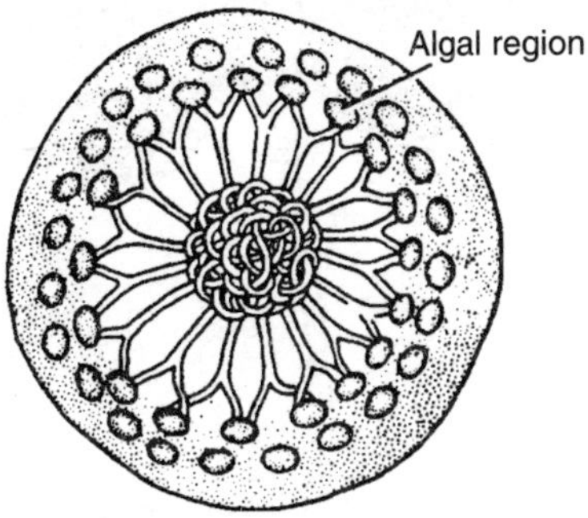

Fig. 2.29. *Usnea sp.* A fruticose thallus. Transverse section to show internal structure

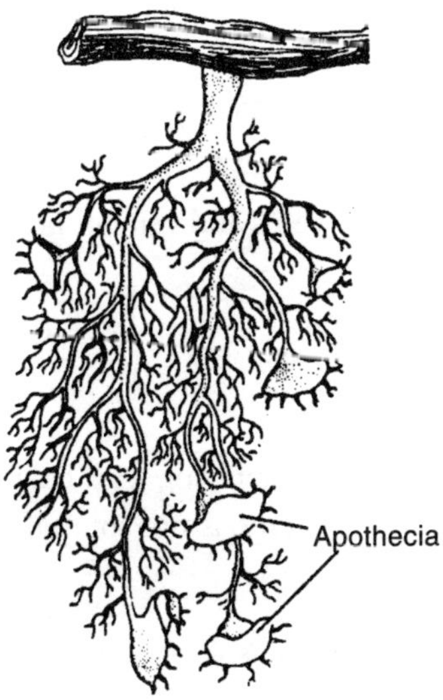

Fig. 2.30. *Usnea sp.* A fruticose ascolichen to show external features

Fig. 2.31. *Cladonia flabelliformis.* A fruticose ascolichen showing external features

2.9 PLANT VIRAL DISEASES

(i) *Name of the disease:* *Bhendi Vein Clearing*
Name of the host: ***Abelmoschus esculentus***
Pathogen: *Yellow vein mosaic virus*

The disease of bhendi is the most serious handicap in the cultivation of the crop in India.

If the plants are infected within 35 days of germination their growth is retarded. Few leaves and fruits are formed and the loss may be about 94%.

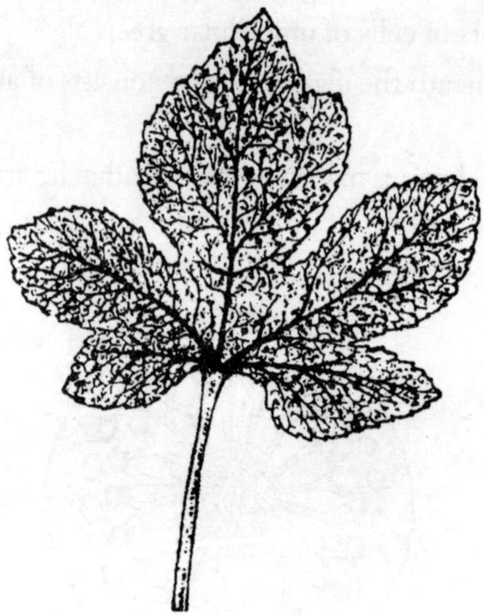

Fig. 2.32 Yellow Vein Mosaic or Vein Clearing of Bhendi

Disease Cycle

The virus is transmitted by white fly and it is not sap transmisable.

The main symptoms are

1. The yellow network of veins is very conspicuous and the veins and veinlets are thicked due to chlorosis of tissue along the veins.
2. Fruits are normal yellowish green in colour.

Control Measures

Protection of the crop from white fly and other insect vectors by spraying with Follidol- 0.3% and other suitable insecticides reduce the incidence of the disease. If the crop is not sprayed within 20 days of germination the disease incidence remains high.

(ii) Name of disease: Tobacco Mosaic Virus
Name of the host: Nicotiana tabacum
Name of the pathogen: Nicotiana virus I.

Symptoms

1. The first symptom to appear is clearing of veins on the young leaves.
2. The leaves develop characteristic light and dark green patterns all over the lamina.
3. The leaf spots may also appear on leaves.

Fig. 2.33 Tobacco mosaic. Leaf of tobacco showing symptoms

4. The diseased plants are usually reduced in size and appear as stunted plants.
5. In severe infection the leaf become narrow, thin and malformed.
6. The disease causes partial sterility to the pollen.
7. The virus is sap transmissible and enters the host through wounds.
8. The virus is rod shaped measuring 300 nm in length and 18nm in diameter. The molecular weight of the particle is 40million Daltons.
9. The capside is made of 2130 identical proteins subunits called capsomeres.
10. The capsomeres are closely packed and arranged in a regular helix around a central hole of 40 Å diameter. Each protein subunit is made up of 158 amino acids.

11. Each capsomere has a furrow on the inner side and the furrows form a continuous helical groove.
12. The genome of the virus is single stranded RNA made up of 6500 base pairs.

2.10 MYCOPLASMAS

Mycoplasmas are named due to the filamentous nature (myco = fungus). Colonies of mycoplasmas exhibit a characteristic fried egg appearance. Mycoplasmas are the smallest free-living organisms, which can pass through bacterial filters. They are pleomorphic—branched, beaded, swollen, ringed or stellate. They lack a cell wall and delimited by three-layered plasma membrane. They are either parasitic/ saprophytic, aerobic / anaerobic and are gram negative. The genome is made up of 500 to 1,100 base pairs of DNA.

Name of the disease: Little Leaf of Brinjal
Name of the host: Solanum melongena
Name of the pathogen: Mycoplasmas

Symptoms

1. The main symptom of the disease is production of very short leaves.
2. The petioles become very much reduced and the leaves very close to the branches and stem.
3. Infected leaves are narrow, soft, smooth and yellowish in colour. New leaves also become reduced in size.
4. The internodes are very much shortened and the axillary buds are stimulated to grow into short branches with small leaves. This gives the plant a bushy habit.
5. The plants do not form flowers and if they are formed they remain green.
6. Fruits are formed rarely.

The disease is transmitted by the vector *Cestius phycitis*.

Control Measures

1. Antibiotic treatment with tetracycline.
2. Eradication of weed host and diseased plant from the field.
3. Control of insect vectors by insecticide.

2.11 BACTERIA—GRAM STAINING

A gram stain is a differential staining method, developed by Henry Christian Gram, a Danish physician in the year 1884, that is why it is called gram staining.

It is very useful for identifying bacteria into two major groups (1) Gram positive (2) Gram negative.

In this process fixed bacteria smear is subjected to four different reagents in the order listed.

1. Crystal violet (primary stain)

2. Iodine Solution
3. Alcohol (Decolouring agent)
4. Saffranin (A counter stain)

The colour changes occur in bacterial cell at each stage due to the four reagents in the gram staining process

The bacteria which retain the primary stain i.e., dark blue or violent or not decolourized when stained with gram method are called gram +ve whereas those that loose the colour violet and counter stain with saffranin appears pink are called gram –ve.

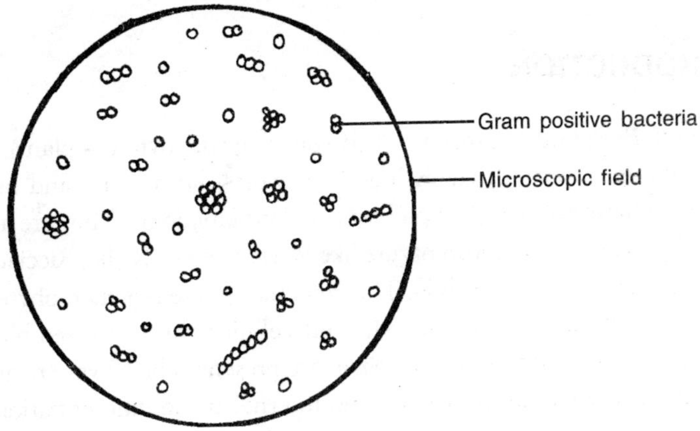

Fig. 2.34 Gram staining of bacteria

Procedure

1. Make a thin smear on the glass slide from the culture medium.
2. Let the smear dry.
3. Heat the smear.
4. Hold the smear using slide rack or cloth pin.
5. Cover each smear with crystal violet for 30 seconds.
6. Wash each slide with distilled water for 30 seconds using wash bottle.
7. Cover each smear with iodine with 95% alcohol drop by drop until no more colour flows from the smear.
8. Gram +ve bacteria are not affected but Gram –ve bacteria are completely decolourised..
9. Wash the slide with distilled water for 30 seconds and drain.
10. Apply saffranin for 30 seconds.
11. Wash with the distilled water and blot dry by blotting paper.
12. Let the slide air dry.
13. Examine the slide under microscope i.e., under 100x using oil immersion.

CHAPTER 3

BRYOPHYTES

3.1 INTRODUCTION

The term *Bryophyta*, a Greek word (Bryon = moss, phyton = plant) is used as a collective name to represent a group of plants that include liver worts, horn worts and mosses growing in amphibious environment. These are called "amphibians of plant kingdom". They are a small group of most primitive land dwellers. A few are aquatic in nature like *Riccia fluitane,* Riella, Ricciacaepus nataae. The plant body is differentiated into stem and leaves. The dominant phase is gametophyte. The gametophyte is fixed to the substratum by means of branched, multicellular rhizoids resembling roots Vascular tissue are completely absent. In folios forms, the leaves are present. The leaves are arranged in 2–3 rows and there is no mid-rib. In thalloid forms, there is no internal tissue differentiation in the thallium. Bryophytes multiply by means of vegetative reproduction and form special structures such as tuber, gaemmae etc.,

Sexual reproduction is by advanced oogamous type. Male gametes, the antherozoid are motile and small, female gametes, the eggs are large and non-motile. Antheridia are all club-shaped and produce many unicellular uninucleate antherozoids. Archegonia are flask shaped with a venter at the back and neck above. Fertilization takes place in presence of water. The oospore develop into a sporophyte. The sporophyte comprise foot, seta and capsule. The capsule contains spore mother cells and on reduction division produce tetrads. The spores germinate and develop into gametophytic plant.

3.2 MARCHANTIA

Vegetative Thallus

Division — Bryophyta
Class — Hepaticopsida
Order — Marchantiales
Family — Marchantiaceae
Genus — *Marchantia*

1. Plant body is gametophytic, thalloid, dorsiventral and prostrate.
2. Thallus is dichotomously branched and the apex of each branch is notched. The growing point occurs at the notch.
3. The upper surface is dark green and fleshy. It has a conspicuous midrib with a shallow longitudinal groove.

4. The dorsal surface shows polygonal areas called areolae. Each polygonal area represents underlying air chamber and has a central air pore.
5. Certain cup like structures are present along the midrib and these are known as gemma cups. These contain gemmae which are the vegetative reproductive bodies. The upper margin of the cup are fringed.
6. Ventral surface of the thallus is brown in colour. It has a low median ridge on the ventral side.
7. The thallus bears rhizoids and scales. Rhizoids are either smooth walled or tuberculate type.
8. Scales are arranged in two to four longitudinal rows on either side of median ventral ridge. They are multi-cellular, one cell thick, wedge shaped and violet in colour. The scales are of two types apendiculate or ligulate, appendaged scales are bigger and have three parts namely neck, body and appendage. Ligulate scales are ligule like.

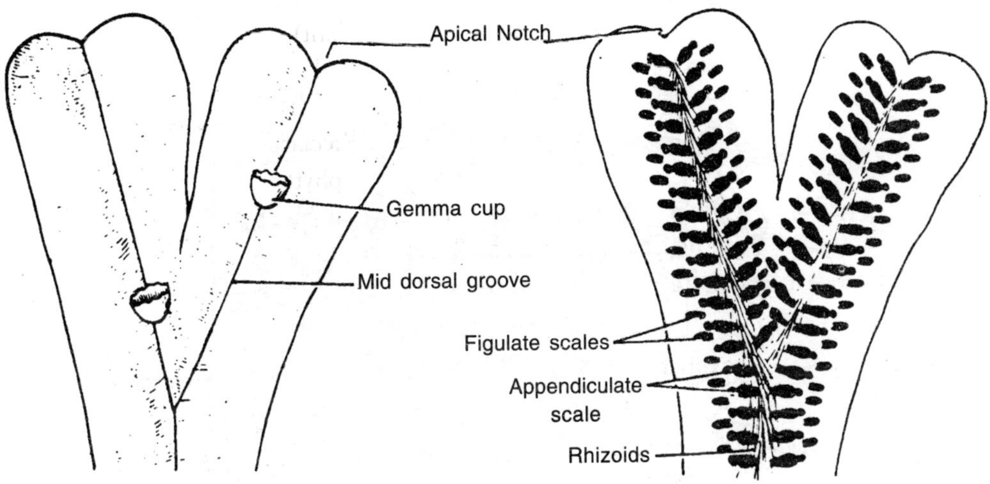

Fig. 3.1

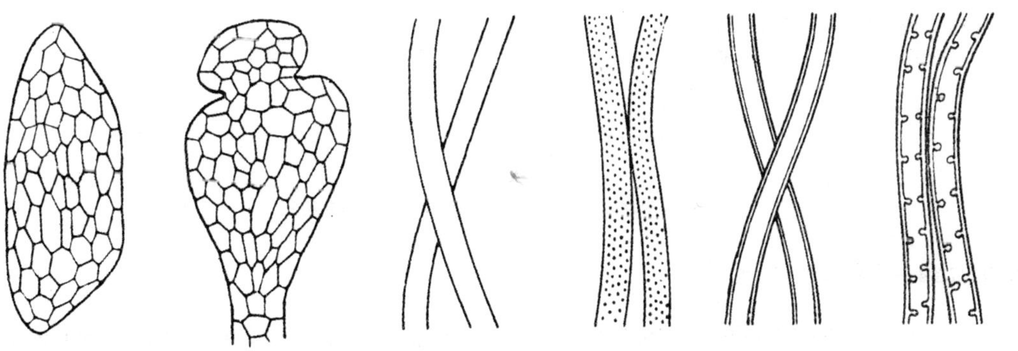

Fig. 3.2 *Marchantia.* External features: dorsal and ventral views, scales and rhizoids

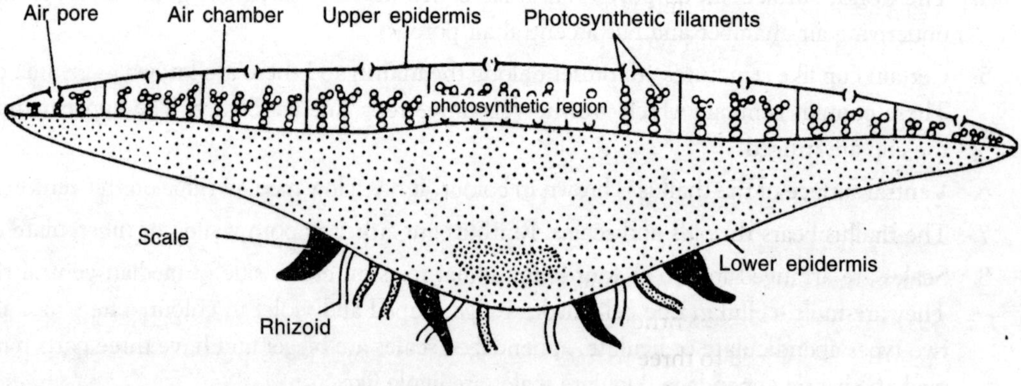

Fig. 3.3 *Marchantia.* V.T.S. thallus (diagrammatic)

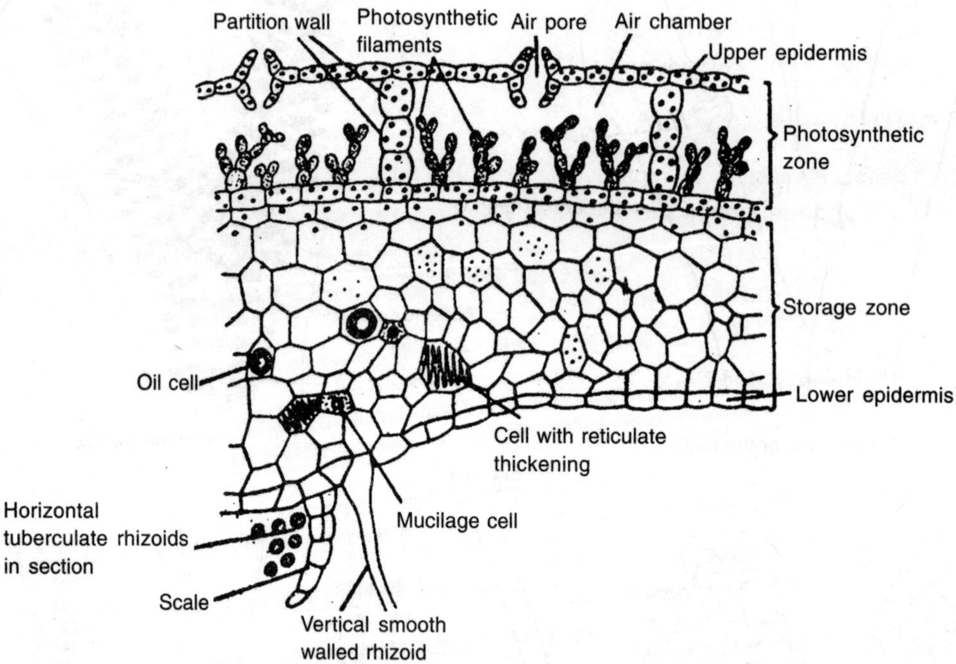

Fig. 3.4 *Marchantia.* Section through thallus

Marchantia—T.S. Thallus

1. Thallus of *Marchantia* is differentiated into an upper photosynthetic, assimilatory region and storage region.
2. The photosynthetic region is divided into many air chambers by septa. It is bounded on the upper side, by the upper epidermis, which consists of a single layer of thin walled cells with slightly thickened outer walls.

3. Embedded in the epidermis are special chimney like or barrel shaped air pores. Superimposed tiers of four or five cells surround each pore.
4. The opening of the pore is slightly raised above the surface of the thallus. Each pore leads into an underlying chamber.
5. The presence of air pores facilitates gaseous exchange necessary for photosynthesis and respiration. These pores are analogous to the stomata of higher plants.
6. Air chambers are arranged in a simple horizontal layer and are filled with assimilatory filaments which posses number of chloroplasts.
7. Just below the photosynthetic region lies the ventral storage region. It is thickest in the middle and reduced to three to four layers of cells towards the margins. It consists of relatively large, colourless, thin walled, polygonal, parenchymatous cells, which are compactly arranged. Some cells contain reserve food materials like starch, protein grains, oil bodies, mucilage and reticulate thickening.
8. Lowermost layer of this region is considered as lower epidermis, from which smooth walled and tuberculate rhizoids and scales arise which are for anchorage and absorption.

MARCHANTIA—Gemma W.M.

1. Gemma is vegetatively reproducing structures formed in the gemma cups that are situated on the mid dorsal line of the thallus.
2. Each gemma is a biconvex lens shaped small, green, multi-cellular structure.
3. It is several cells thick in the middle portion and deeply notched on the two opposite edge.
4. In each marginal notch growing point is present.
5. Majority of the cells in the gemma are green due to abundant chloroplasts. A few cells contain oil bodies and a few cells are colourless and known as rhizoidal cells.
6. Each gemma on germination produces two thalli because from each notch develops a single individual thallus.
7. It is a very rapid devise for multiplication.

MARCHANTIA—V.S. of Gemma Cup Through Thallus

1. The gemma cup arises as a part of the thallus. Its base remains attached to the thallus.
2. In vertical section the cup appears like a shallow cavity. On the floor of the cup are present gemmae.
3. Structurally, margin of the cup is similar to thallus differentiated into photosynthetic and storage region.
4. Gemma is vegetatively reproducing structures formed in the gemma cups that are situated on the mid dorsal line of the thallus.
5. Each gemma is a biconvex, lens shaped, small, green multicellular structure.
6. It is several cells thick in the middle and deeply notched on the two opposite edges.
7. In each marginal notch, growing point is present.

8. Majority of the cells in the gemma are green due to abundant chloroplasts. A few cells contain oil bodies and a few cells are colourless and known as rhizodal cells.
9. Each gemma on germination produces two thalli and from each notch a single individual thallus develops.
10. It is a very rapid devise for multiplication.

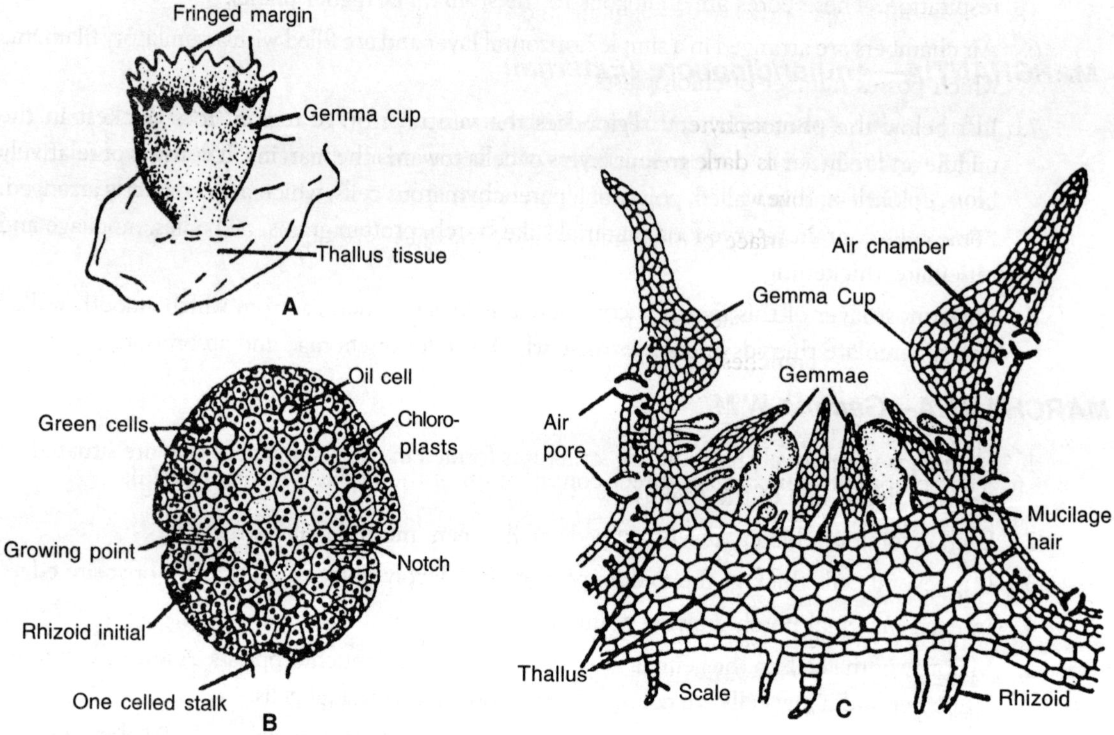

Fig. 3.5 *Marchantia*: **A.** Gemma cup entire, **B.** Gemma cup showing structure, **C.** Vertical transverse section of the thallus lobe passing through the gemma cup

MARCHANTIA—Vegetative Thallus with Gemma Cups

1. Plant body is a gametophyte, thalloid, dorsiventral and prostrate. It is dichotomously branched. The upper surface is dark green, fleshy with conspicuous dorsal midrib, which is a shallow longitudinal groove.
2. The dorsal surface shows polygonal areas called aerolae. Each polygonal area represents underlying air chambers and has a central air-pore.
3. Certain cup like structures are present along the midrib and these are known as gemma cups.They have fringed margins.
4. Each cup encloses a number of gemmae, which are vegetatively reproducing bodies.

5. Each gemma is biconvex, disc like, small green, multicellular structure attached with a unicellular stalk to the floor of the cup.
6. From the lower surface of the thallus along the mid ventral ridge, scales and rhizoids are produced.
7. Rhizoids are of two types—smooth and tuberculate.
8. Scales are again of two types—ligulate and appendaged.

MARCHANTIA—Antheridiophore (External)

1. Plant body is a gametophyte, thalloid, dorsiventral and prostrate. It is dichotomously branched. The upper surface is dark green, fleshy with conspicuous dorsal midrib, which is a shallow longitudinal groove.
2. From the lower surface of the thallus along the mid-ventral ridge, scales and rhizoids are produced.
3. Plant is dioecious—male plant produces antheridiophores.
4. These are the branches of limited growth arising just near the apical notch of the thallus lobes.
5. Each branch has a stalk and terminal horizontal disc.
6. These upright branches are direct continuation of the thallus growing upwards.
7. The cylindrical stalk is one to three cms in height. The disc is eight lobed. Each lobe represents an apex of a branch. Growing points occur at the tip of each lobe. A little away from the growing point the antheridia are produced in acropetalous succession.
8. Stalk has five faces, two faces are curved outwards and three faces are furrowed longitudinally. The dorsal side contains air chambers, photosynthetic cells and airpores like that of the dorsal thallus surface.

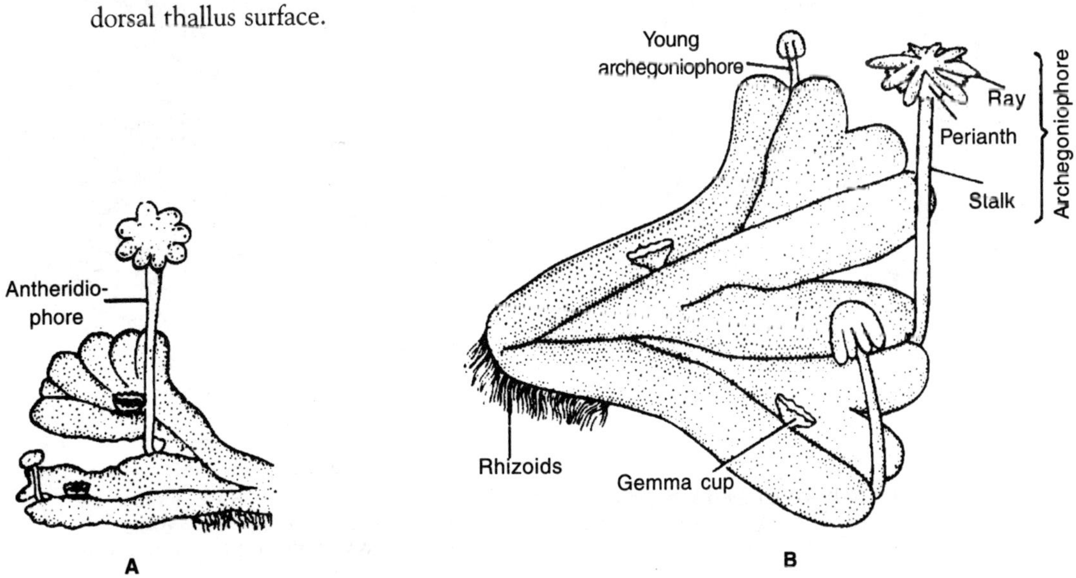

Fig. 3.6 *Marchantia.* **A.** Thallus with antheridiophore, **B.** *Marchantia.* Thallus with archegoniophores

MARCHANTIA—L.S. Antheridiophore

1. It has a cylindrical stalk, which is one to three cms. in height and bears a lobed terminal horizontal disc.
2. The stalk shows dorsiventral symmetry. It shows two rhizoidal grooves on the lower side, situated one on either side. Upper surface has photosynthetic region divided into many air chambers.
3. The peltate disc is slightly convex and the internal structure resembles that of the thallus.
4. Alternating with the air chambers are antheridial cavities. Each cavity is flask shaped and opens to the exterior by an ostiole.
5. Each cavity bears single antheridia. These antheridia are produced in acropetal succession of eight lobes.
6. Each antheridium is a stalked body covered by a single layer of jacket cells and encloses androcytes inside. These androcytes metamorphose into biflagellate antherozoids.

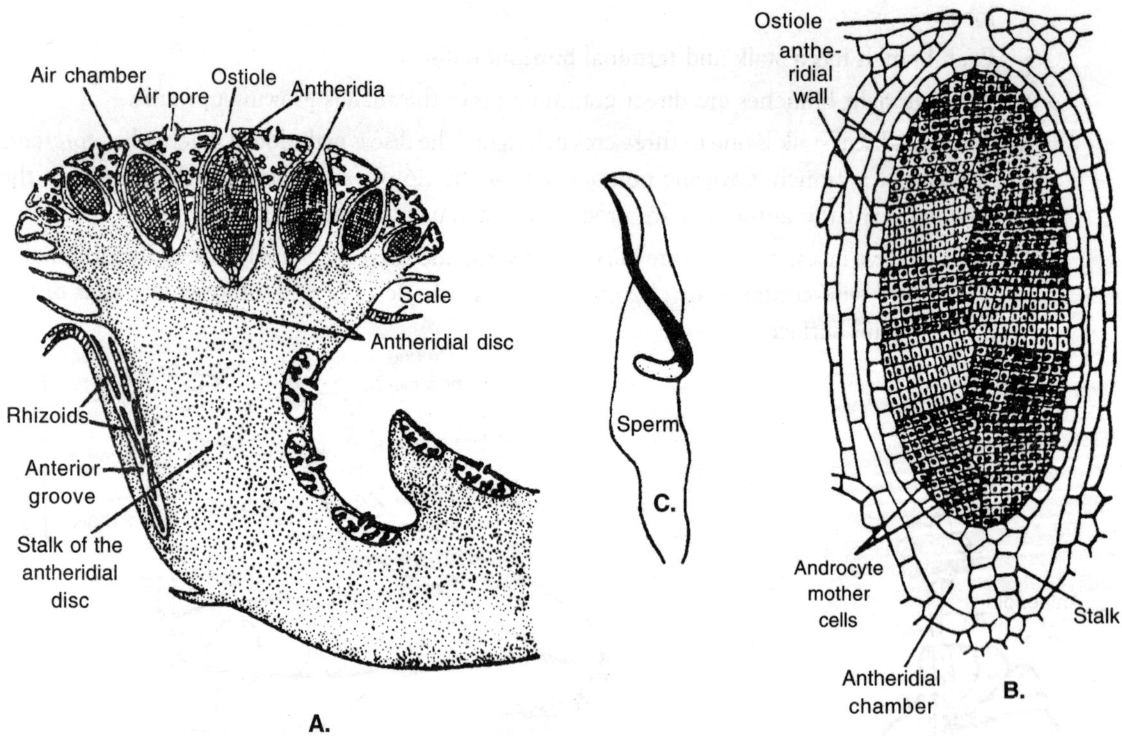

Fig. 3.7 *Marchantia.* A longitudinal section through the young antheridiophore; B.V.S. through the antheridial chamber containing a nearly mature antheridium C, biflagellate antherozoid

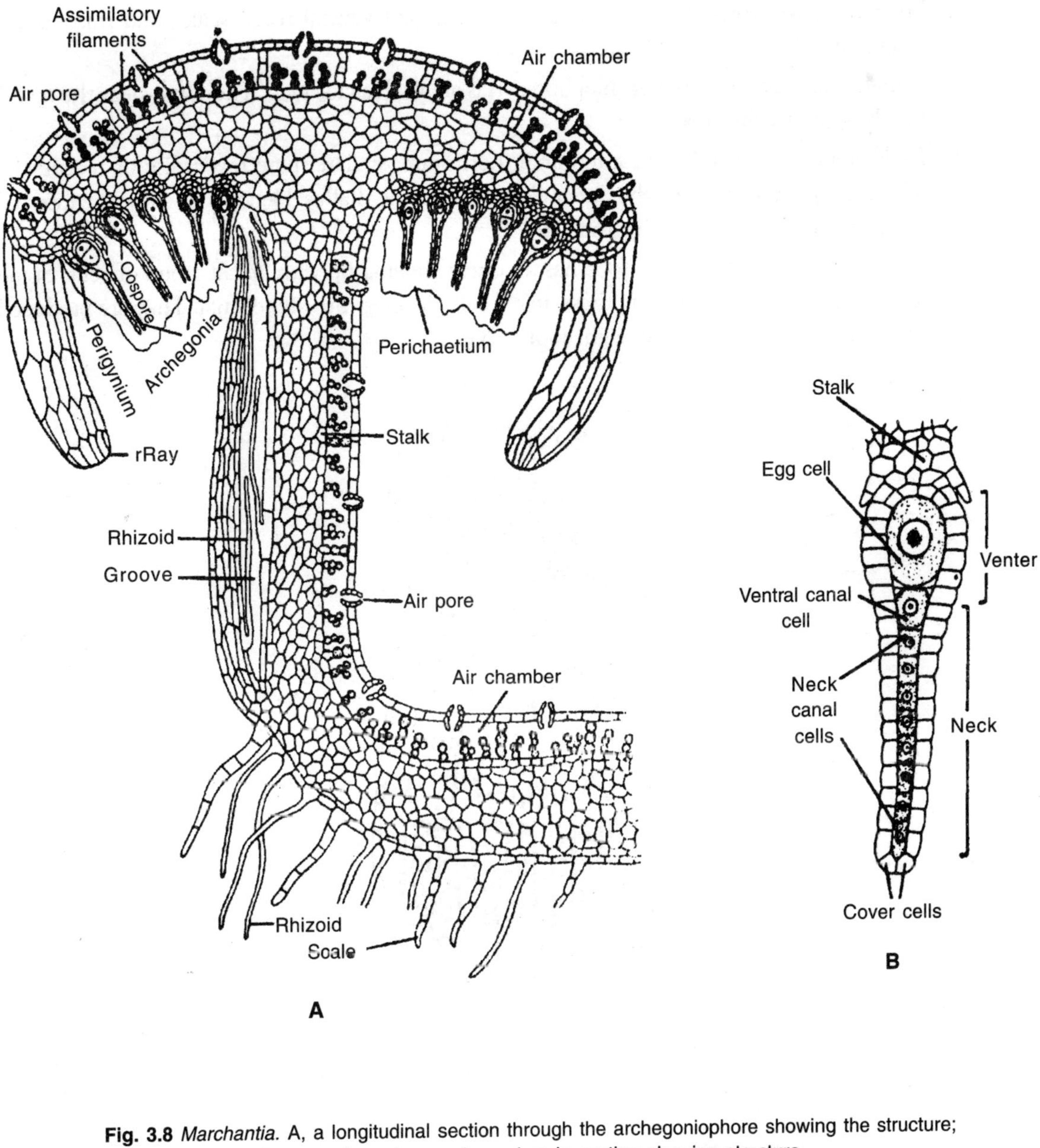

Fig. 3.8 *Marchantia.* A, a longitudinal section through the archegoniophore showing the structure; B, a nearly mature archegonium in section showing structure

MARCHANTIA—Archegoniophore (External)

1. Plant body is a gametophyte, thalloid, dorsiventral and prostrate. It is dichotomously branched. The upper surface is dark green, fleshy with conspicuous dorsal midrib, which is a shallow longitudinal groove.
2. From the lower surface of the thallus along the mid ventral ridge, scales and rhizoids are produced.
3. Female plants produce archegoniophores. It is about five cms long. They arise near the apical notch of the thallus lobes.
4. Each branch has a stalk and a terminal, horizontal, stellate, nine rayed disc or receptacle. These upright branches are the direct elongation of the thallus. Hence they show the internal structure similar to vegetative thallus.
5. The disc is eight lobed. Each lobe has an outer growing tip with apical group of cells. Some dorsal segments given out from these apical cells develop into archegonia. They are produced in acropetalous succession in eight radial rows.

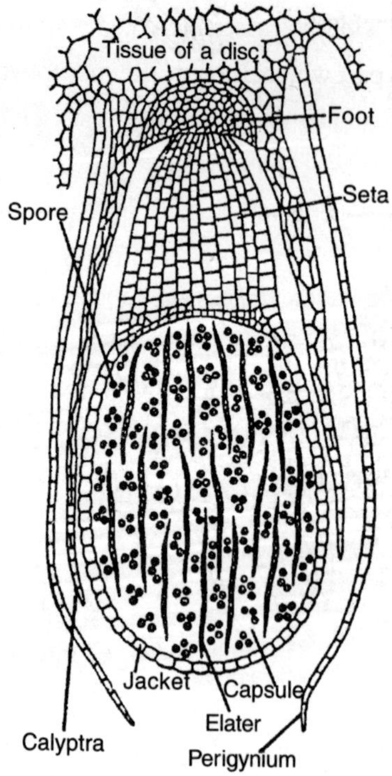

Fig. 3.9 *Marchantia.* L. S. mature sporogonium

6. After fertilization the archegonia are pushed towards the lower surface of the disc.
7. Long, stout finger like processes grow out from the periphery of the disc between group of archegonia. These are called rays. Usually they are eight in number. They radiate outwards and bend from inwards as an umbrella ribs and give a stellate appearance at maturity.

MARCHANTIA Sporophyte — L.S.

1. The sporophyte develops from the archegonium after its fertilization. Therefore capsules are seen in a disc of mature archegoniophore on the lower side. Only one sporophyte develops in one involucre.
2. The sporophyte is enclosed by three coverings.
   (*i*) Calyptra
   (*ii*) Perignium (perianth)
   (*iii*) Perichaetium (involucre)
3. Calyptra forms a protective covering around the yound sporophyte.
4. Repeated division of basal cell of archegonium produces a one-celled thick collar like outgrowth outside the calyptra. It is called perigynium.
5. The mature sporophyte is differentiated into foot, seta and capsule.
6. Foot is the basal bulbous part which anchors the sporophyte to the disc of the archegoniophore.
7. Seta is a short and stout stalk that connects foot to the capsule.
8. The mature capsule is a yellow coloured oval structure.
9. Capsule has a single layered jacket, inside which lie many spores and elaters. Spores are arranged in tetrahedral tetrads.
10. The spores are round and have a diameter of about 12-30 μm.
11. A spore has an outer thick sculptured exine and a thin uniform intine. Every spore is uninucleate with rich cytoplasm.
12. Elaters are spindle - shaped and posses two spiral thickening bands. These are hygroscopic and help in the dispersal of spores.

3.3 ANTHOCEROS (Horn Worts)

Division — Bryophyta
Class — Anthocerotopsida
Order — Anthocerotales
Family — Anthocerotaceae
Genus — *Anthoceros*

1. Anthoceros is world wide in its distribution. The genus is represented by 200 sps.
2. It grows in shady places as dense plates.
3. The plant body is a gametophyte, which is small, yellowish green or dark green dorsi ventrally flattened, lobed thallus. The mid-rib is absent.

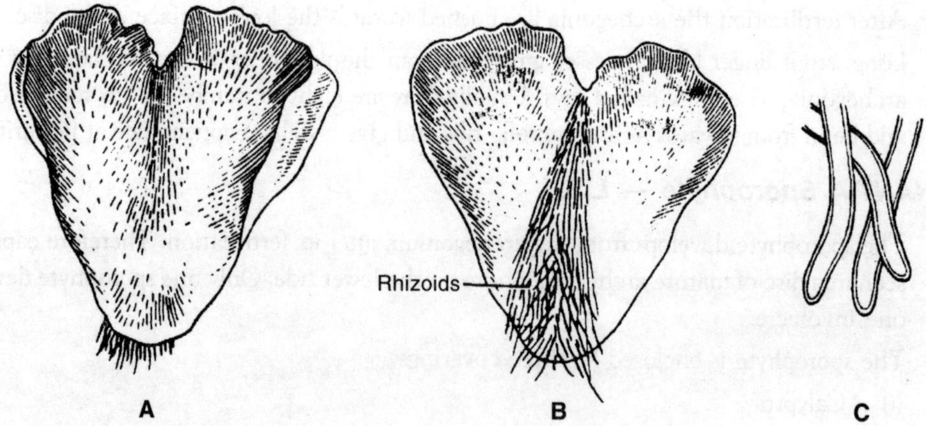

Fig. 3.10 *Anthoceros.* External features of gametophyte. **A.** Dorsal surface. **B.** Ventral surface. **C.** Rhizoids

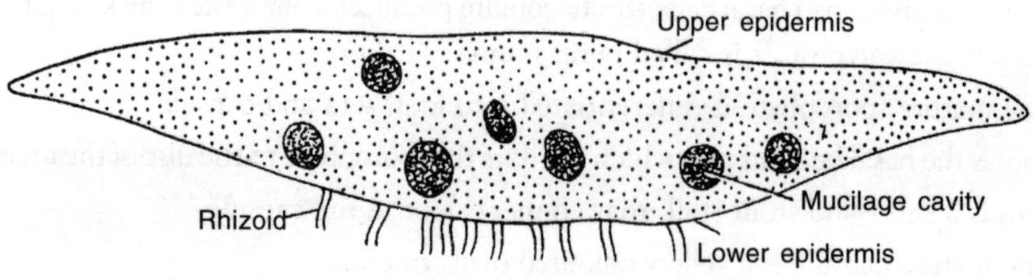

Fig. 3.11 *Anthoceros.* V.t.s. thallus (diagrammatic)

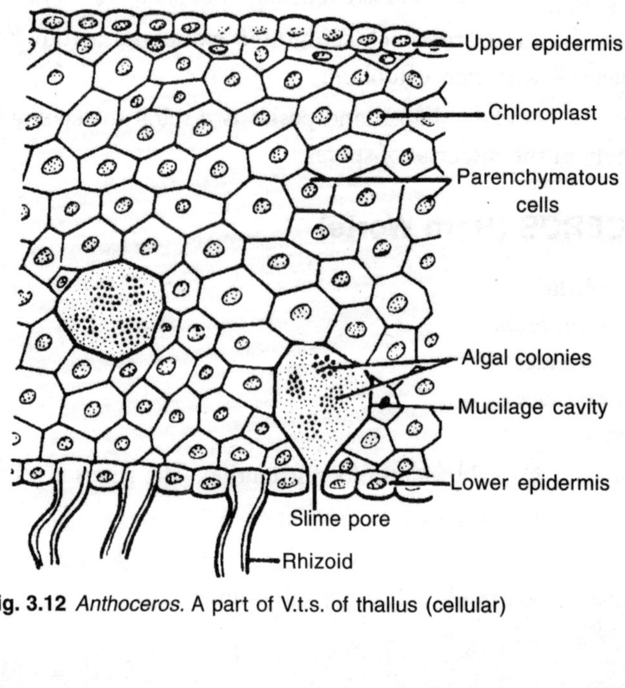

Fig. 3.12 *Anthoceros.* A part of V.t.s. of thallus (cellular)

4. The dorsal surface of the thalluss is smooth, velvetty or rough.
5. The ventral surface bears numerous unicellular smooth walled rhizoids. They anchor the thallus to the substratum.
6. On the ventral surface of the thallus, several small, bluish green spots are present, which correspond to the mucilage cavities that contain endophytic *atga Nostoc*.

ANTHOCEROS—T.S. Thallus with Nostoc

1. The thallus of *Anthoceros* is very simple in organization and cellular differentiation. All the cells are nearly alike.
2. Internally the thallus is made of several layers of cells. Mid rib is absent. The entire thallus consists of soft parenchyma, which is compactly arranged.
3. There is no organized epidermis.
4. Each cell of the thallus contains a large chloroplast with single pyrenoid, except few layer of epidermal cells which give rise to rhizoids.
5. On the ventral side there are certain mucilage filled cavities, which open by some pores called slime pores. They are guarded by guard cells. They are functionless and do not control the size of the pore, which thus remains completely open.
6. In these mucilaginous cavities *Nostoc* colonies are present. They live in symbiotic association with thallus. They fix atmospheric nitrogen for the thallus and inturn take organic carbon from it.
7. From the lower epidermis in middle region single celled smooth walled rhizoids are given.

ANTHCEROS—Antheridia V.S.

1. Antheridia in *Anthoceros* are produced in acropetal succession, on the dorsal surface of the thallus.
2. Antheridia are endogenous in origin. They are enclosed in antheridial cavity filled with mucilage. The roof is two layered.
3. They occur singly or in groups. Ripe antheridia are bright orange in colour.
4. Each antheridium has an ovoid or pouch-like body. It has a multicellular slender stalk. It is made of four vertical rows of cells.
5. The body of the antheridium consists of a jacket layer enclosing the androcytes inside.
6. The cells of the jacket layer are arranged in four super imposed tiers. They are rectangular in shape. They also contain one plastid each. When the antheridia are young the plastids are colourless, later on they become green and ultimately bright orange or red.
7. Each androcyte metamorphoses into a biflagellate sperm.

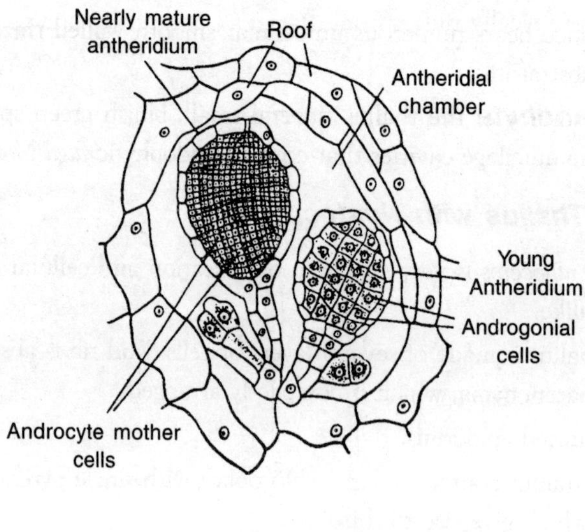

Fig. 3.13 *Marchantia.* Mature antheridia

ANTHOCEROS—Archegonia V.S.

1. The archegonia are sunk deep in the fleshy thallus on its upper surface.
2. They are developed in acropetal order.
3. Each archegonium consists only an axial row of four to six neck canal cells, a ventral canal cell and an egg.

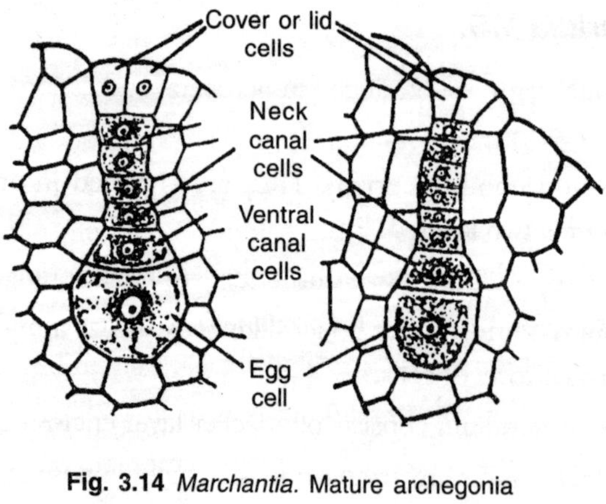

Fig. 3.14 *Marchantia.* Mature archegonia

4. There is no sterile jacket layer except the distal rosette of cover cells forming its tip.
5. The cover cells slightly project above the general, upper surface of the thallus where it is usually surrounded by a funnel shaped mass of mucilage called the mucilage mound.
6. The surrounding vegetative cells of the thallus offer protection to the cells of the axial row.

7. The archegonium of *Anthoceros is* immersed in the thallus and is in direct contact with the surrounding vegetative cells, differs from that of all Bryophytes and resembles certain of the Pteridophytes.

ANTHOCEROS—Gametophyte with Sporophyte

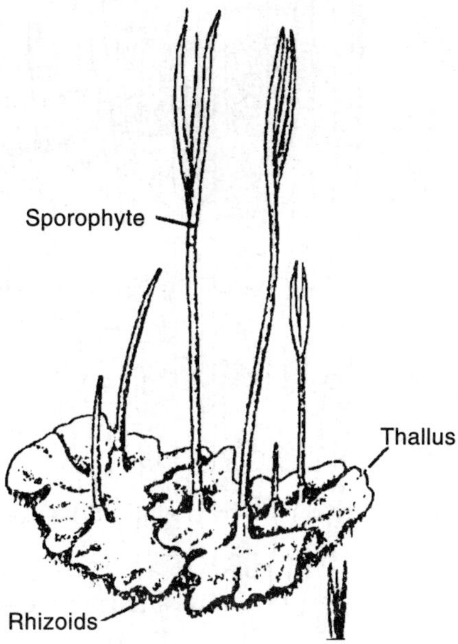

Fig. 3.15 *Anthoceros.* Gametophyte with Sporophyte

1. Plant body is a gametophyte, dark green, dorsiventral and prostrate. It shows wavy and lobed margins. Midrib is not distinct. Margin is lobed. They show dichotomous type of branching. It is several cells in thickness.
2. Ventral surface of thallus bears numerous smooth walled unicellular rhizoids. It also shows small dark opaque spots where bluish green colonies of *Nostoc* are formed.
3. From the upper surface of the thallus long, cylindrical, delicate sporangia are produced. They arise in clusters. Each sporangium has a tubular sheath around at its base. It is the involucre.
4. It is differentiated into foot, meristematic zone and the capsule.
5. Foot is bulbous and remains embedded in the gametophytic tissue.
6. Meristematic zone helps in elongation of the capsule by gradual addition of cells.
7. Capsule is long upright structure reaching 2–15 cm.
8. It produces meiospores. At maturity it opens by two valves exposing the central columella. The two valves curve outward give the shape of horn. Therefore the name 'Horn-worts'.

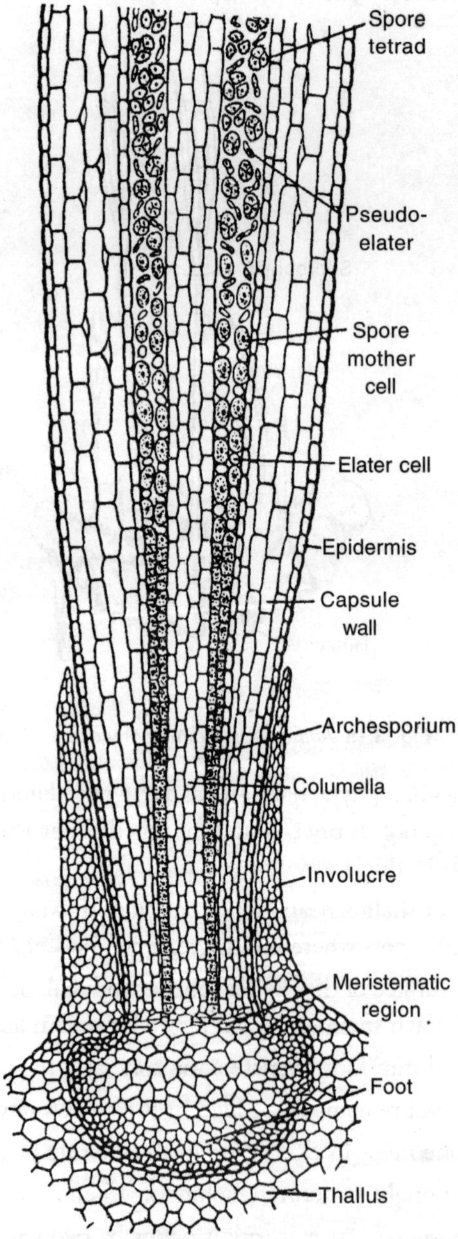

Fig. 3.16 *Anthoceros.* L. S. through the sporogonium

ANTHCEROS—Sporophyte - Spore Region

1. The sporophyte of *Anthoceros* is differentiated into three regions. Foot, intercalary zone and capsule. The seta is absent and its place is taken by intercalary zone.
2. The size of the sporogonium ranges from one to several centimeters. At the base a tubular sheath called involucre surrounds it. It is gametophytic structure and protective in function.
3. Foot—It is a rounded bulbous structure, deeply embedded in the tissue of the gametophytic thallus. It is haustorial in function.
4. Meristematic zone—It is a narrow zone of meristematic cells located at the base of the capsule just above the foot.
5. It constantly adds new cells to the capsule at its base. They become progressively differentiated into columella, archesporium and capsule wall. The presence of basal intercalary meristematic zone enables the capsule to grow for a long period and form spores.
6. Capsule—It is long slender, smooth, cylindrical and upright (2–15 cm). In the center of the capsule is present columella, which is 4–16 cells in thickness. It is endothecial in origin.
7. Next to the columella is the sporongenous tissue, which develops from the inner layer of the epithecium. Sporogenous tissue is one cell thick in the beginning but at maturity it become two to four layered. Slightly above in the capsule the sporogenous tissue is differentiated into spore mother cells and pseudoelater mother cells arranged in alternate bands.
8. At slightly higher-level spore tetrads and spores as well as many elongated multicellular pseudoelaters are present.
9. The capsule is bounded by four to six layered capsule wall. Outermost layer is the epidermis. It contains a few stomata. Each cell of the capsule contains two chloroplasts.

3.7 POLYTRICHUM (Pigeon Wheat Moss)

Division — Bryophyta
Class — Bryopsida
S.cl — Bryidae
Order — Polytrichales
Family — Polytrichaceae
Genus — *Polytrichum*

The genus *Polytrichum* comprises about 100 sps. They grow on wide variety of habitats which include clayey soil, dry stony places, dry wood, sandy barks, swamps and some on wet soils.

1. The plant body is a gametophyte which is differentiated into an underground rhizome and aerial, erect, leafy stem.
2. The rhizoids are produced by rhizome, and they are long thick walled with oblique septa. The rhizoids coil round one another to form a rope like structure. These strands provide mechanical support.

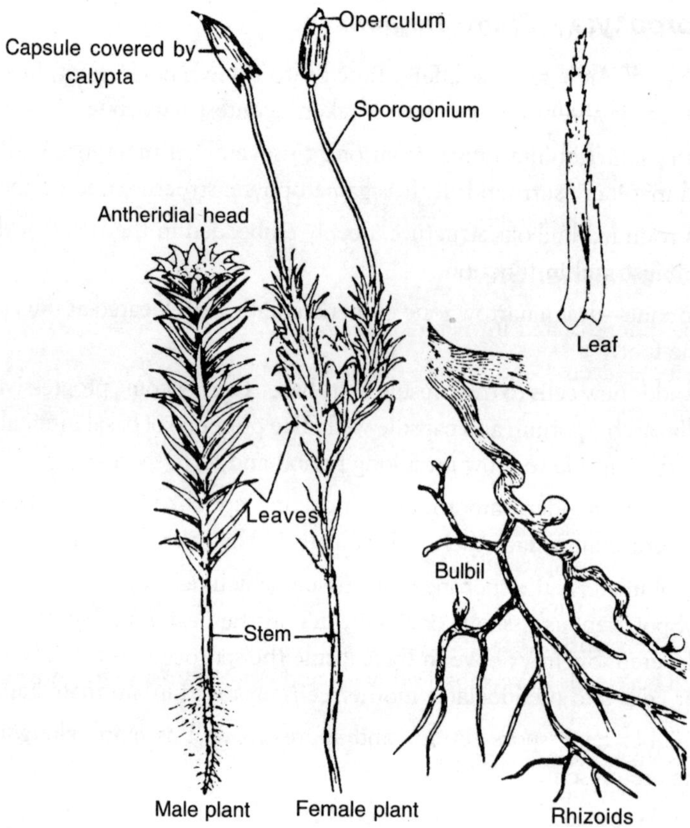

Fig. 3.17 *Polytrichum.* External features of gametophore: male and female plant, leaf and rhizoids

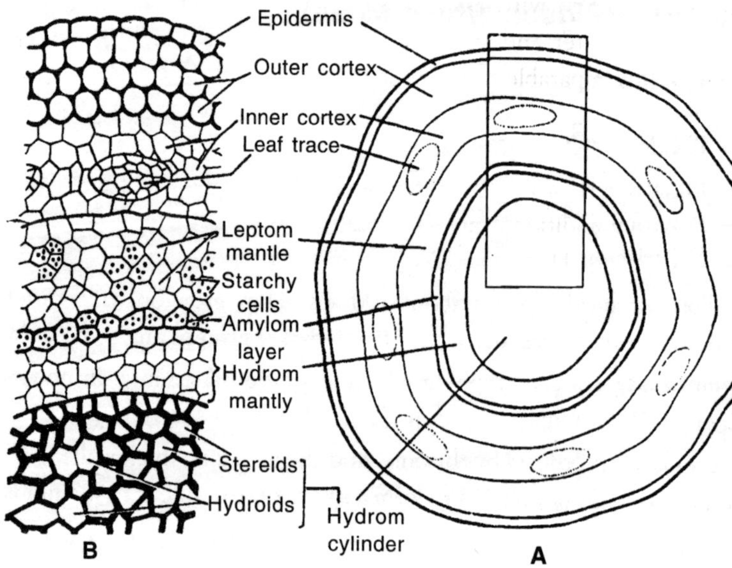

Fig. 3.18 *(A–B)Polytrichum.* **A.** T.S. aerial stem (outline sketch), **B.** portion of A showing details

3. The aerial shoot grows to a height of about 20 cms. The leaves on the aerial stem are spirally arranged. They are either green, brown or colourless.
4. The leaves on the rhizome and middle transitional region are in three vertical rows. These are brown in colour or colourless. Each leaf possesses a broad, colourless, membranous, one celled sheathing leaf base which narrows into a limb. The margins of the wings are coarsely toothed. The leaf has a dark green midrib.

POLYTRICHUM—T.S. of Aerial Stem

1. A stem can be differentiated into epidermis, cortex and central cylinder.
2. Epidermis is single layered and usually not well defined.
3. Within the epidermis is a wide cortex. It is differentiated into outer thick walled, prosenchymatous cells and inner thin walled parenchymatous cells. Inner cortex contains chloroplasts. Leaf traces are present in this region.
4. Central cylinder consists of central core of compactly arranged thick walled cells called hydrome cylinder. It comprises of steroids and hydroids.
5. External to hydrome cylinder is a zone consisting of two to three layers of thin walled cells called as hydrome mantle. External to hydrom mantle is a narrow zone consisting of single layer of cells. These cells have suberised walls and store starch—Hydrom sheath or amylom layer.
6. Just external to the amylom layer is a broad irregular zone composed of thin walled. Sieve tube like cells called as leptom mantle.
7. A layer of parenchymatous cells, delimits cortex from the central cylinder and is called as endodermis or starchy sheath (as the cells of this layer contain starch).
8. Leptom is concerned with translocation of assimilates and hydrom is concerned with translocation of water and minerals. It is said that the well-developed central tissues of *Polytrichum* are comparable to the protosteles of lower pteridophytes (*Rhynia* and *Psilotum*).

POLYTRICHUM—Leaf T.S.

1. Each leaf consists of two parts sheathing leaf base and a distal limb. The *Polytrichum* leaf is unique in having additional photosynthetic tissue in the form of thin, closely set, vertical green plates of cells. They are one cell layer thick and 5–8 cells high.

   In cross section the limb consists of a broad nerve or mid rib. Lamina is rudimentary and represented by rudimentary wing. It is comprised of single layer of cells.
2. The wide mid rib is multilayered in the middle and gradually thins towards the margins.
3. Major part of this contains thin walled parenchymatous cells. Interspearsed among them are patches of thick walled cells.
4. The lower epidermis is well marked.
5. Upper surface is bounded by a layer of large thin walled cells from which lamellae arise.
6. The terminal cells of the lamellae are bifid.

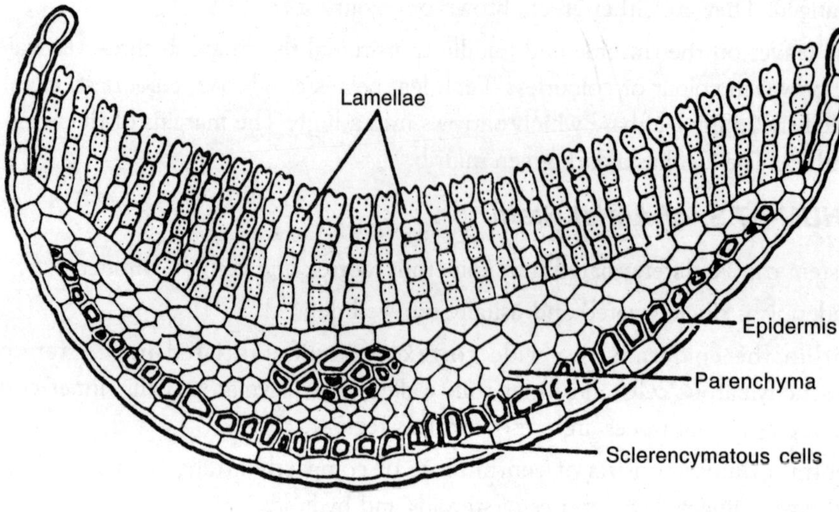

Fig. 3.19 *Polytrichum.* T.S. Leaf

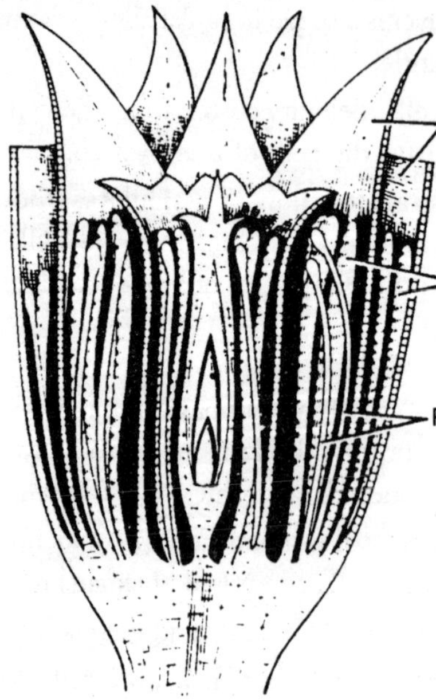

Fig. 3.20 *Polytrichum.* L.S. through antheridial head

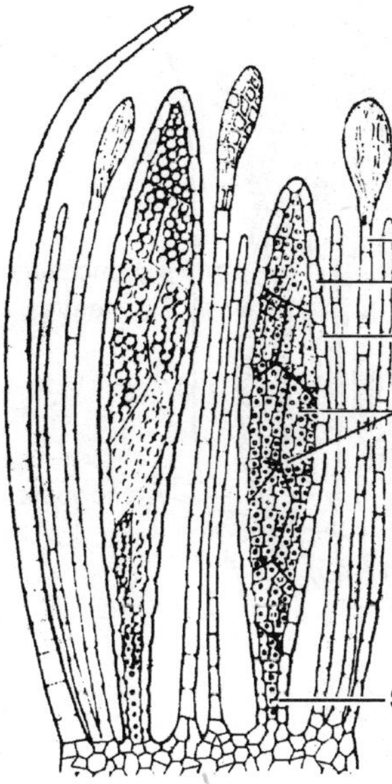

Fig. 3.21 *Polytrichum.* Antheridia and paraphyses

POLYTRICHUM – V.S. of Antheridial Head

1. Polytrichum is dioecious and male plants produce antheridial clusters from below the perigonial leaves. The antheridial head is also called as male flower as it superficially resembles miniature angiospermic flower.
2. The antheridia are surrounded by specialized leaves known as perigonial leaves.
3. The perigonial leaves are short, broadly expanded at the base terminating in a short bristle.
4. The antheridia are produced in groups at the base of each perigonial leaf in the position of lateral bud.
5. Intermingled with antheridia are the paraphyses. Some of these are filamentous, others are broadened at their tips into a spathulate single-celled thick mass of cells.
6. Each antheridium is a stalked, club shaped body, 1–6 mm. in length. It is covered by a single layer of jacket enclosing androcytes.
7. Since the apical cell is not used up in the formation of antheridia, growth of the male shoot is not arrested.
8. At maturity antheridia dehisces and spermatozoids are released.

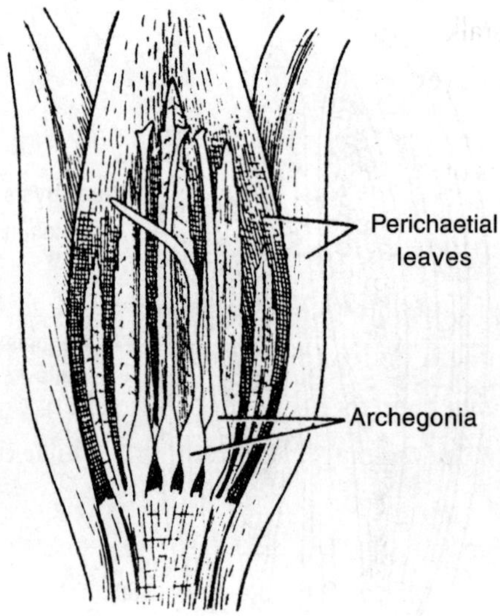

Fig. 3.22 *Polytrichum.* L.S. through archegonial head

POLYTRICHUM—V.S. of Archegonial Head

1. The archegonia occur in a group at the apex of the female gametophore.
2. The leaves surrounding the archegonia are called perichaetial leaves.
3. Intermingled with the archegonia are paraphysis which are filamentous.
4. The apical cell is used up in the formation of archegonia.

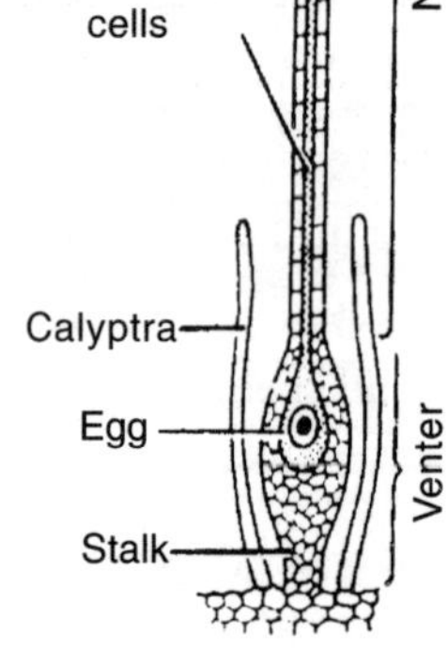

Fig. 3.23 *Polytrichum.* An archegonium

5. Each archegonium is a typical flask shaped structure consisting of very long neck, broad venter and massive stalk.
6. The venter wall is two layered and neck is made up of six vertical rows of cells forming the wall of the neck.
7. The axial row consists of egg cell, venter canal cell and 13 neck canal cells.

POLYTRICHUM—Gametophyte with Sporophyte

1. The dominant phase in the life cycle is gametophyte. It grows to a height of 25–30 cms.
2. It is differentiated, into an underground rhizome and aerial, erect leafy stems.
3. In *Polytrichum communae* there is a transitional zone between rhizome and upper leafy shoot.
4. Rhizoids are borne on the rhizome, they are multi-cellular thick walled and characterized by oblique septa.

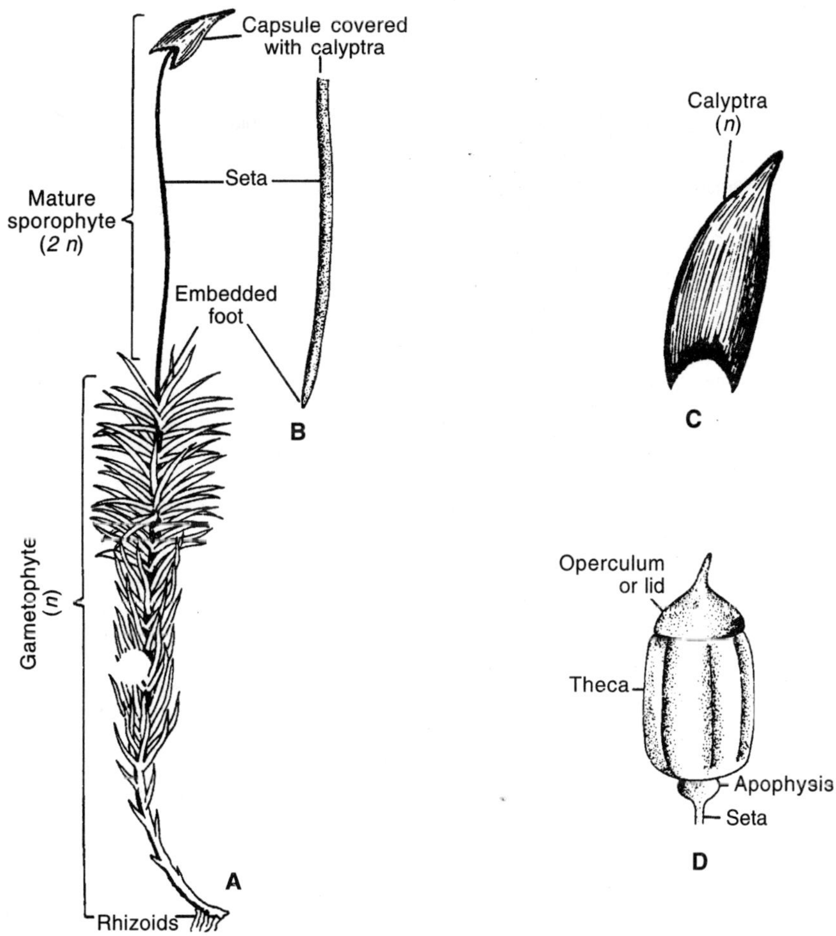

Fig. 3.24 *Polytrichum* **A.** Female gametophore bearing terminally a nearly mature sporophyte, **B.** Lower portion of seta ending in a dagger-shaped foot, **C.** calyptra removed from the capsule, **D.** mature capsule

5. Several such rhizoids twist and coil around each other forming a cord. It retains water in capillaries and also gives mechanical support to the plant.
6. The leaves are arranged spirally on rhizome. They are in three vertical rows and show colours of brown on aerial shoot. They are large green, 3/8 arrangement.
7. Each leaf has a broad membranous colourless sheathing leaf base and narrow lanceolate limb. The leg has a broad mid rib (Margin is toothed and apex acute).
8. The gametophyte bears sporophyte at its apex, which is differentiated into foot, seta and capsule.

POLYTRICHUM—Capsule L.S.

1. The sporophyte of *Polytrichum* is differentiated into foot, seta and capsule.
2. Foot is dagger shaped structure with parallel sides and tapering tips. It is buried deep in female gametophore to which the sporophyte is conically attached.

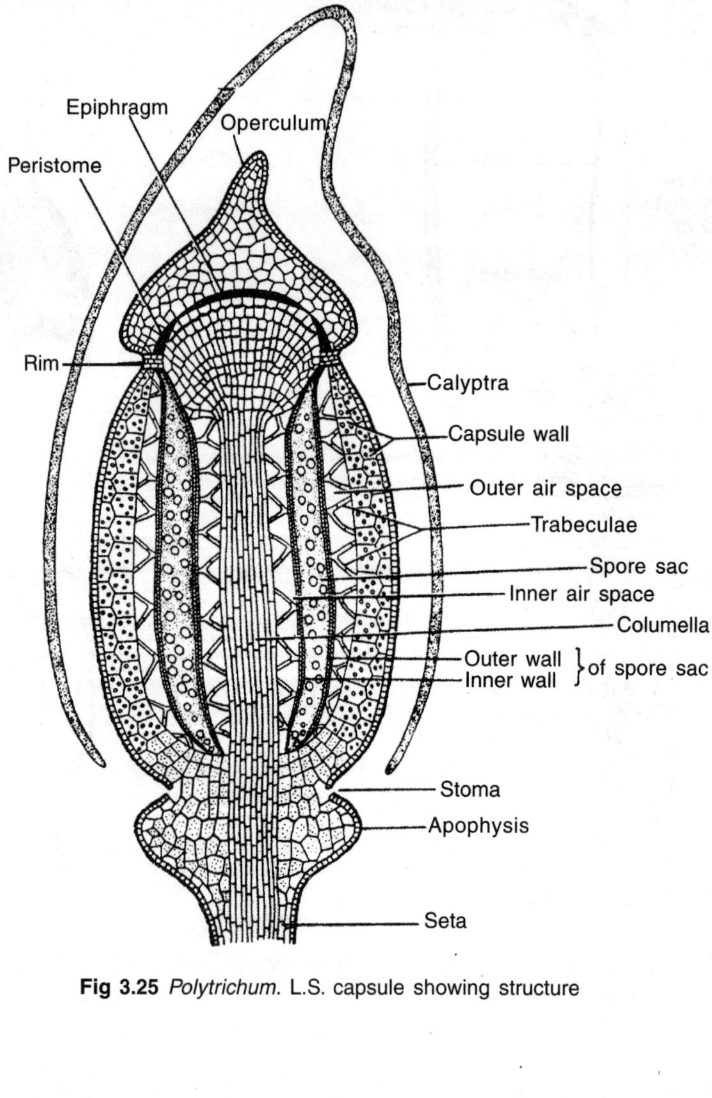

Fig 3.25 *Polytrichum.* L.S. capsule showing structure

3. Seta is long, slender stalk like structure about 5 cm long and 0.5 mm in diameter. Seta is differentiated into epidermis, hypodermis, cortex and central hydroid cells.
   Support and conduction are its two functions.
4. Just below the capsule the seta is considerably swollen to form bulbous, sterile structure called the apophysis. It is specialised for photosynthesis and has stomata in the grooves.
5. At the tip of seta is capsule. It is erect in position and almost completely covered by brown, hairy calyptra.
6. The capsule is differentiated into two regions–
   (*i*) The lower, spore bearing angular portion called *theca*
   (*ii*) A sterile conical portion forming the lid or operculum.
7. The theca has several layered thick capsule wall (A layer of epidermis and two layers of chlorophyllous tissue). Within the wall is outer air space traversed by trabeculae connecting the capsule wall with outer spore sac wall. Next to the outer air space is the spore sac consisting of a large number of meiospores. Inner to the spore sac is inner airspace traversed by trabecular which connect the inner spore sac wall to columella. The outer and inner spore sac wall is two layered columella is made up of vertically elongated cells and this is concerned with conduction.
8. Upper end of theca joins with the operculum at narrow constriction made of 2–3 layers of radially elongated cells forming rim.
9. The columella expands into operculum in the form of a fan called epiphragm. It is thin, membranous, shield shaped structure.
10. The operculum is a conical portion of the capsule which fits like a cap at the mouth of the theca region and is drawn out into a rostrum at its distal end.
11. A row of 32 or 64 peristomial teeth develop below the epiphragm.

CHAPTER 4

PTERIDOPHYTA

4.1 INTRODUCTION

The Pteridophyta (term derived from the Greek words Ptecon = feather and Phyta = plants, means plants with feather like founds of ferns), incudes a group of primitive land plants with well developed vascular system. This group has also been refered to as Vascular Cryptogams and classified by Carolus Linnaeus (1754) under the class Cryptogamia.

The Pteridophytes grow under varied habitats. Most of them are terrestrial plants and thrive well under damp and shady conditions, while some flourish well in open grasslands exposed dry places or even under xeric conditions. A few Pteridophytes grow as aquatic, semiaquatic or epiphytic plants.

The main independent plant body of Pteridophytes is sporophyte. It is differentiated into true roots, leaves and stems, only primitive members lack true roots and well developed leaves. Plants vary in size and form. They may be small annuals (eg. *Azolla* and *Salvinia*) or large tree like perennials as *Angiopteris* and *Osmunda*. Leaves may be scale like, small and sessile or large, petiolate and compound.

Vegetative organs possess vascular tissues xylem and phloem but xylem lacks vessels and phloem lacks companion cells. The stele may be a protostele as in *Lycopodium*, or siphonostele as in *Equisetum*, or dictyostele as in *Pteris* or polycyclic as in *Angiopteris*.

Plants may be homosporous or heterosporous. Sporangia produce spores and the development of sporangia may be eusporangiate or teptosporangiate. Spores develop into prothalli on which antheridia and archegonia develop. In aquatic forms like *Marsilea*, the sporangia are found inside specialized structures called sporocarps. The development of gametophyte may be exosporic or endosporic. The antheridia produce antherozoids which are unicellular, uninucleate, flagellated and motile. Archegonium has a neck usually with four vertical rows of cells and two neck canal cells but some species have more than two or only one also. Water is always needed for the act of fertilization and also for completion of life cycle. From zygote, sporophyte is developed. The plants exhibit heteromorphic type of alternation of generation.

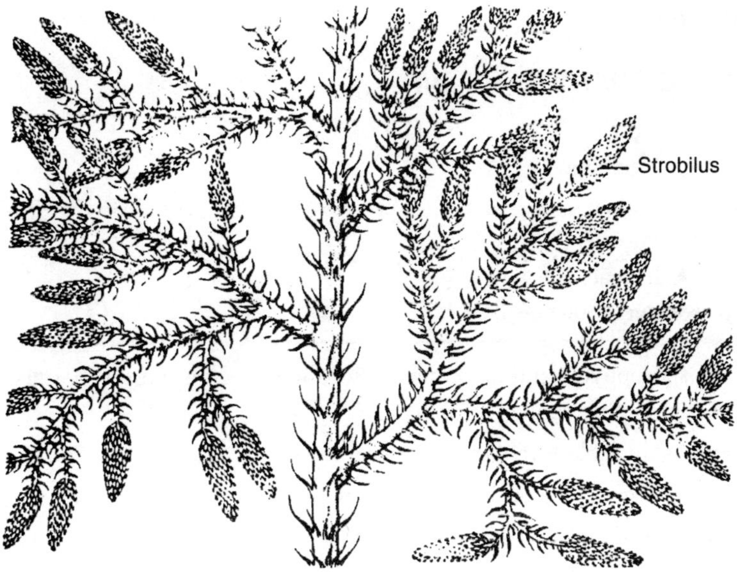

Fig. 4.1 *Lycopodium cernuum* sporophyte with strobilus

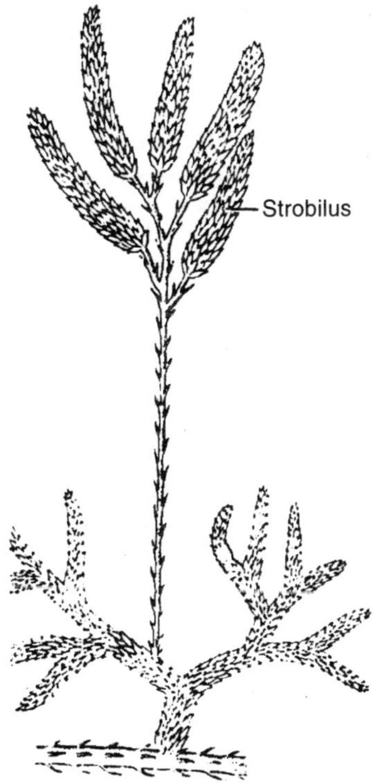

Fig. 4.2 *Lycopodium clavatum* sporophyte with strobilus

4.2 LYCOPODIUM

LYCOPODIUM SPOROPHYTE—Habit

Division — Pteridophyta
Sub-division — Lycophyta
Class — Euligulopsida
Order — Lycopodiales
Family — Lycopodiaceae
Genus — *Lycopodium*

1. Plants are sporophytes and are commonly called as 'club mosses' because of their moss like appearance and club shaped strobili.
2. Plant body is divisible into stem, root and leaves.
3. Plants are evergreen herbs and consist of horizontally creeping rhizome, which gives off slender elongated aerial branches from upper side and adventitious roots from the lower surface.
4. Branching in the first formed portion is dichotomous but in later formed portion it appears monopodial due to unequal dichotomy and is called pseudomonopodial.
5. The stem and its branches are densely covered with small simple sessile leaves.
6. All the leaves are alike in shape and size, i.e., isophyllous and are arranged in whorls, or in spirals.
7. Leaves are microphyllous provided with a small mid vein, which does not run up to the apex.
8. The first formed root is short lived and adventitious roots arise endogenously from the under surface of the creeping stem in an acropetalous manner.
9. At the apex of elongated stalk like branches strobili are present which are homosporous.
10. Each strobilus has a central axis, which bears numerous spirally arranged sporophylls having stalks. On their adaxial side close to the axis single kidney shaped sporangia are present.

LYCOPODIUM CERNUUM—T.S. of stem

Anatomy of stem: It consists of three regions. Epidermis, Cortex and Stele.

1. **Epidermis:** It is single layered and provided with stomata. Outer walls of the epidermal cells are thick and cutinized. The stem is having certain ridges and grooves. In the ridges, chlorenchyma is present.
2. **Cortex:** It is divided into outer and inner thin walled cortex and middle sclerenchymatous cortex.
3. **Endodermis:** It is last layer of inner cortex. It is clearly recognizable in the earlier stages of development because of the presence of casparian strips. As the development proceeds in mature stems, the endodermal cells become thickened.

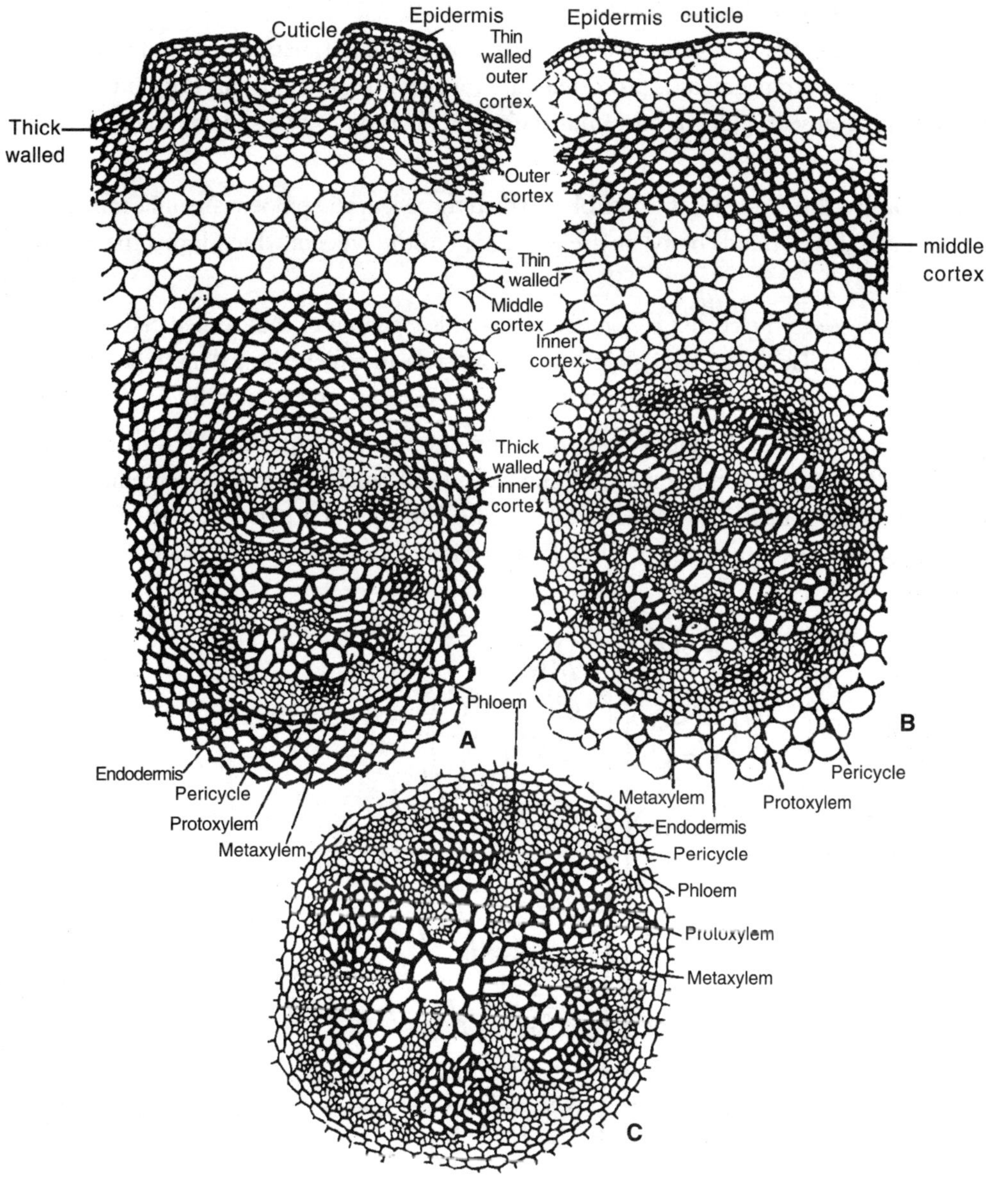

A. *L. clavatum* — T.S. portion of stem. Note inner sclarenchymatous cortex and plectostelic protostele.
B. *L. cernuum* — The middle cortex is sclerenchymatous. The stele is a mixed protostele.
C. *L. serratum* — Actinostelic protostele.

Fig. 4.3 *Lycopodium.* Stem anatomy

4. **Pericycle:** It is present next to the endodermis and is composed of one or many layers of compactly arranged parenchymatous cells.
5. **Vascular bundle:** It consists of only primary tissues i.e., primary xylem and primary phloem. Basically the arrangement is protostele. In protostele the xylem forms the central of the stem and pith is absent, phloem surrounds the xylem. This protostele shows variations in different species.

L. cernuum: It is MIXED PROTOSTELE in which the xylem and phloem are uniformly distributed. In T.S it appears as if the strands of xylem are embedded in the phloem which forms the ground mass. Xylem is exarch.

In *Lycopodium selago* the cortex is homogenous and is made up of thin walled cells which enclose small or large intercellular space. The stele is an actinostele in which the xylem appears star-shaped with protoxylem situated at the tips of the star-shaped projections and is exarch.

In *L.clavatum* the cortex is differentiated into three zones. The outer zone has thick-walled cells, the hypodermis. The middle cortex is made of thin walled cells enclosing many intercellular spaces and the inner cortex is again made up of thick walled cells. The stele is a plectostele in which the xylem appears in the form of separate plates of variable sizes with phloem in between them.

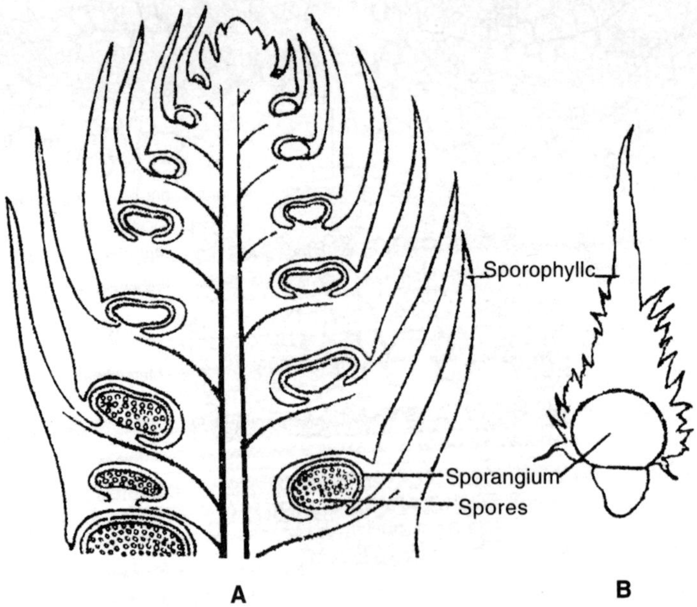

Fig. 4.4 *Lycopodium.* **A.** L.S. apical portion of strobilus. **B.** Dorsal view of sporophyll and a sporangium

LYCOPODIUM—Strobilus (Cone)

1. *Lycopodium* is a sporophyte, it bears spores and reproduces asexually.
2. It is the fertile part of the plant. Strobili are club shaped and are borne in pairs at the apices of elongated stalk like branches called as podia, which are covered with minute scale leaves.

3. Strobili are homosporous. It consists of a central axis on which, large number of acutely pointed, yellow sporophylls are spirally arranged.
4. Each sporophyll bears a single sporangium on its adaxial surface very close to the axis.
5. Sporangium is distinguishable into two prominent parts. (i) stalk and (ii) the capsule.
6. The stalk is short and broad. The capsule is curved and reniform and unilocular.
7. All the sporangia are alike and produce only one kind of spores hence it is homosporous.
8. The sporangium wall is three layered and the spores are arranged in tetrads.

They show characteristic triradiate ridge.

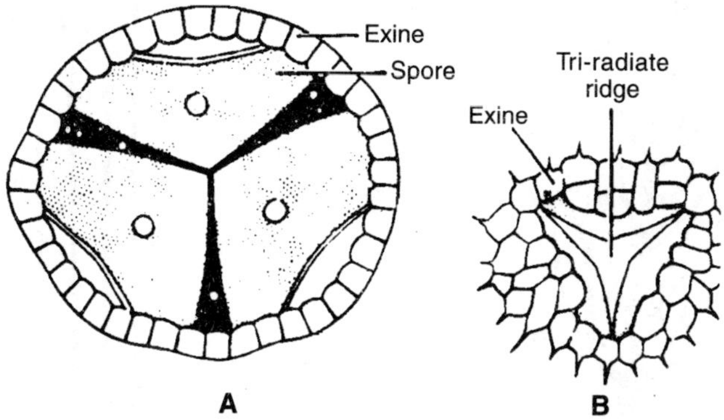

Fig. 4.5 *Lycopodium.* **A.** Spore tetrad, **B.** A spore

LYCOPODIUM—L.S of Cone

1. *Lycopodium* is a homosporous sporophyte.
2. It reproduces asexually by means of spores, which are present in the sporangia.
3. The sporangia are borne on the sporophylls adaxially.
4. The L.S of strobili shows a central axis encircled by spirally arranged compact sporangia bearing sporophylls.
5. Each sporophyll is a small sporangium bearing structure.
6. The sporangium may be axillary or foliar.
7. Each sporangium is kidney shaped orange to yellow structure and shows multicellular stalk and capsule.
8. The wall of the capsule is three layered. The innermost layer is tapetum.
9. Inner to the tapetum are the spore mother cells, which undergo reduction division to produce tetrahedral spore tetrads.
10. Each spore has a triradiate ridge and a rounded semicircular base.
11. The spore wall is two layered outer lamellar and the inner granular.
12. The spores on germination produce gametophyte.

4.3 EQUISETUM - SPOROPHITE

Division — Sphenophyta
Class — Calamopsida
Order — Equisetales
Family — Equisetaceae
Genus — *Equisetum debile*

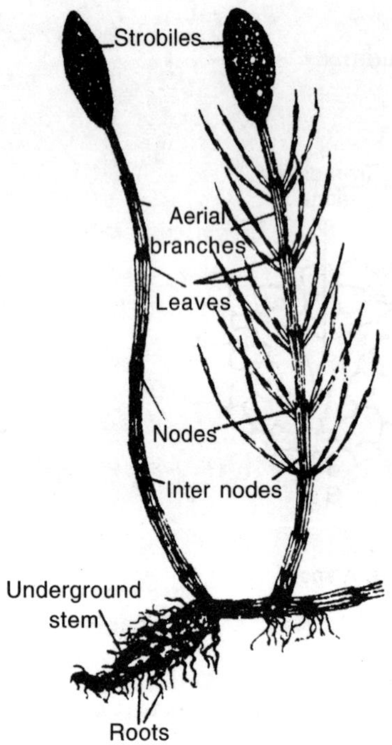

Fig 4.6 *Equisetum sporophyte plant body*

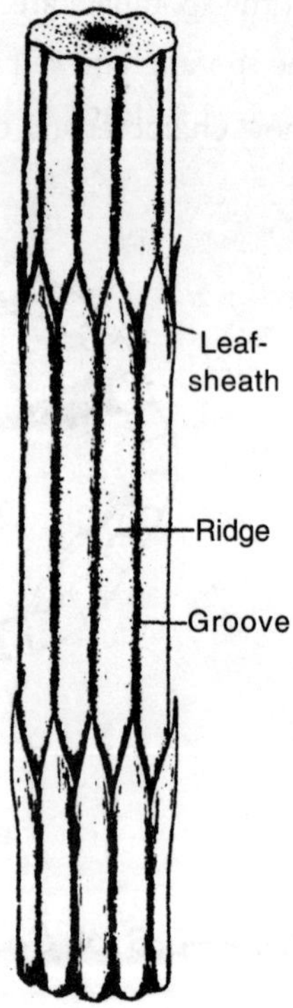

Fig. 4.7 *Equisetum debile.* Portion of an aerial stem showing ridges and grooves and their position in the respective internodes

1. Plant body of *Equisetum* is a sporophyte. It is commonly called as 'horsetail'. It is a well-branched perennial herb.
2. The plants are found in damp and sandy soil. The sporophyte is a perennial herbaceous plant having rhizomatous stem with aerial shoots, adventitious roots and scale leaves.
3. Generally the plants are less than a metre in height with few exceptions. *E.debile* is 1–2 feet high under xeric conditions but grows upto 2–5 metres tall under moist and shaded environments.
4. The stem is articulated, fistular and shows the distinction of nodes and internodes with ridges and furrows and the ridges of successive internodes alternate with one another.
5. The plant body consists of a long, horizontal, underground rhizome, which is sympodially

branched. From this many roots arise and grow towards the lower side. From upper and internodes.

6. Underground rhizome which is long, creeping, well branched is divisible into nodes and internodes.
7. The roots, which develop from the lower side of rhizome, are long, slender, well branched and adventitious.
8. Aerial shoots arise from the upper part of the rhizome and are jointed, green, rough, stiff and erect. The jointed appearance is due to the presence of distinct nodes and internodes.
9. The nodes bear whorled leaves that are laterally fused at the base.
10. The internodes are long and hollow and are longitudinally ridged and grooved. The ridges alternate with successive grooves.
11. The number of ridges corresponds to the number of leaves at a node.
12. Intercalary meristem is present at the base of the internode.
13. The aerial stems may be branched or unbranched. In E. *debile*, they are unbranched.
14. In E. *debile* there is no distinction of fertile and sterile shoots. All are green and may bear terminal strobili.
15. Leaves are microphyllous. They are simple isophyllous and small in size. They arise in whorls at each node.

EQUISETUM—Stem (T.S. of Internode)

In a transverse section, the stem is wavy in outline due to ridges and grooves.

1. **Epidermis**: It is a single layered tissue of the stem forming the strong outer covering. Their walls are cuticularised and covered with rods or grains of silica. The silica strengthens the cell walls, prevents water loss through epidermis, protects against pathogens and predators and helps in keeping the plant erect.

   The continuity of epidermis is broken by sunken stomata present in each groove. In each stoma the guard cells are covered completely by subsidiary cells, thus giving the appearance of two sets of guard cells.

2. **Cortex**: Below the epidermis is present a well-developed cortex. It can be divided into outer cortex and inner cortex. Outer cortex has two types of cells. (i) Sclerenchymatous cells; (ii) Chlorenchymatous cells.

   (i) Sclerenchymatous Cells: These are thick walled cells that give mechanical strength to the stem and sub-epidermal in position. Below the ridges, they occur in large and heavy groups. It is absent below the stomata.

   (ii) Chlorenchymatous Cells: These are present below the sclerenchyma forming a curved band. They contain large number of chloroplasts. It extends up to epidermis in each furrow where lie the stomata. The cells are elongated and in 3–4 layers (forming a continuous band).

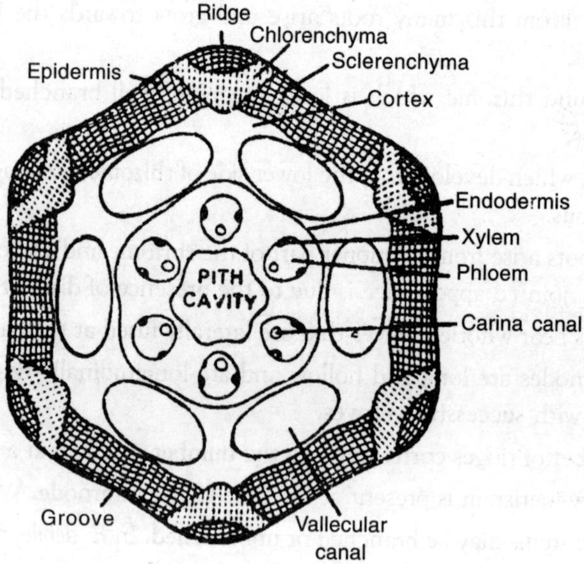

Fig. 4.8 *Equisetum*. T.S. internode of aerial shoot (diagrammatic)

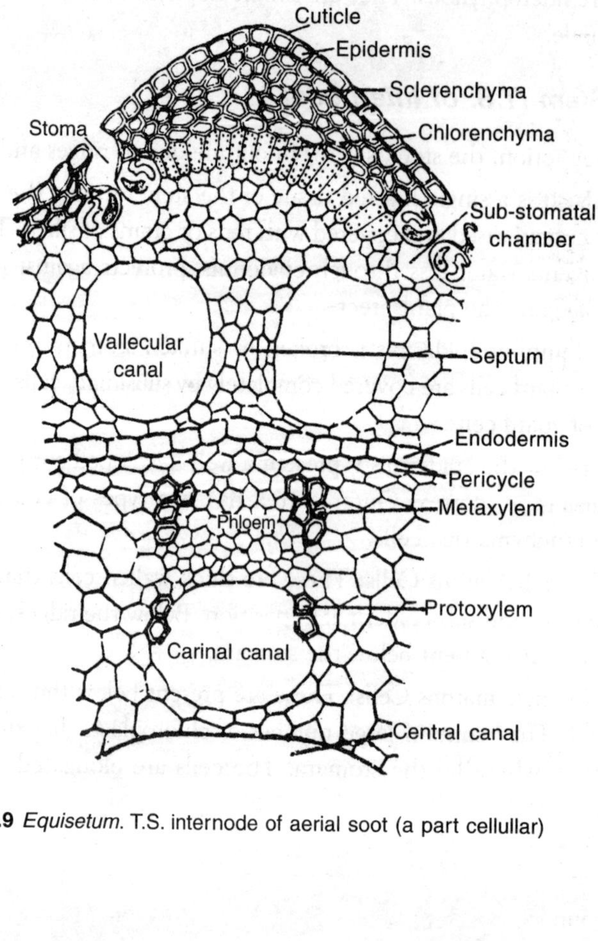

Fig. 4.9 *Equisetum*. T.S. internode of aerial soot (a part cellullar)

Inner Cortex: It consists of a few layers of larger celled parenchyma traversed by a ring of large air spaces forming an aerating system. They are called vallecular canals beneath the photosynthetic tissue.

Endodermis: Innermost layer of the cortex is the endodermis characterized by casparian strips.

Pericycle: There is a single layer of pericycle made up of parenchymatous cells.

Stele: It consists of a well-defined ring of small vascular bundles surrounding a large pith cavity. The bundles are conjoint, collateral and endarch. They alternate in position with the vallecular canals and so lie beneath the ridges. A large zone of parenchyma separates the individual bundles in a ring from one another. Internal to the vascular bundles and beneath the ridges are small carinal canals or cavities. They are filled with water. These canals are formed by the tearing apart of the protoxylem elements.

Vascular bundle consists of xylem and phloem. Xylem is poorly developed. The xylem is in 'V' shape. The point of V lies towards center. The metaxylem forms the two limbs of 'V' and protoxylem occupies the point of 'V'. Later on carinal canals as a result of its disorganization occupy the position of protoxylem.

The phloem consists of phloem parenchyma and sieve tubes. It lies between the two limbs of 'V'.

Pith: it is present in young stem but is absent in mature stem and is represented by large cavity.

EQUISETUM—Strobilus Entire

1. The strobilus or cone of *Equisetum* is developed terminally on the non-chlorophyllous and short-lived shoots known as fertile shoots.
2. Each strobilus has a thick central axis known as strobilus axis.
3. On the strobilus axis are many sporangiophores, arranged at right angles to the cone axis.
4. Each sporangiophore is a stalked structure. The stalk has a shield shaped peltate disc attached to the outer end of the stalk.
5. Many sporangia (5–10) hang from below the disc. They extend towards the axis of the cone and fill up all the space that is left between it and sporangiophores.
6. Each sporangium is elongated and contains a rounded apex.
7. Due to the mutual pressure the peltate discs usually becomes hexagonal in outline.
8. At the base of the cone the central axis bears a ring like outgrowth called annulus. It is a protective structure.
9. Each sporangium is covered by a jacket and encloses haploid spores, which are alike - homosporous.

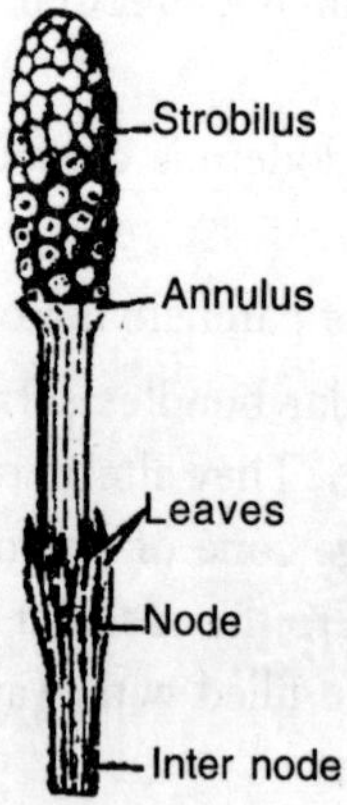

Fig. 4.10 *Equisetum.* Fertile shoot with cone

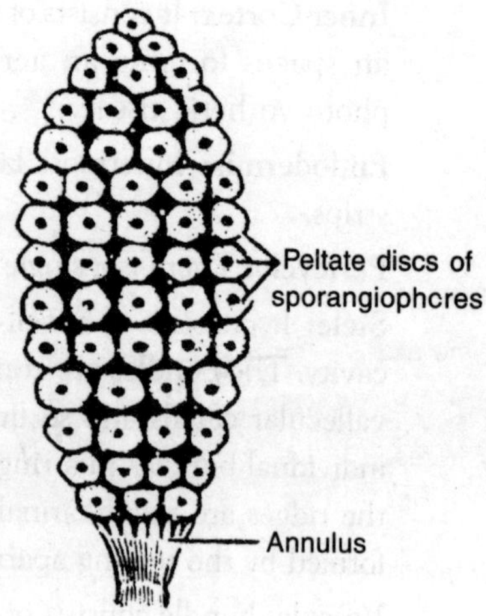

Fig. 4.11 *Equisetum.* A cone

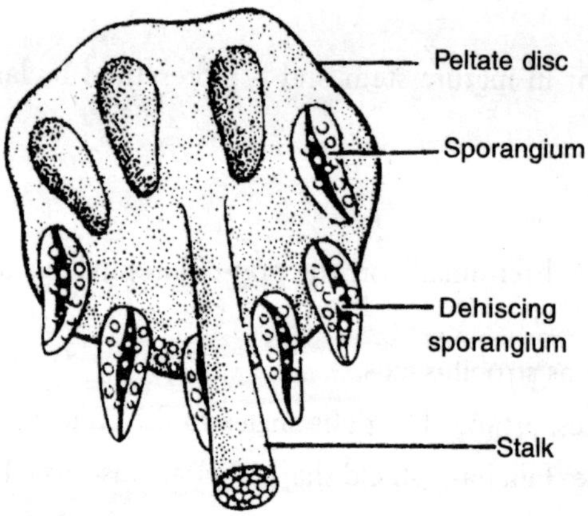

Fig. 4.12 *Equisetum.* Sporangiophore from ventral side

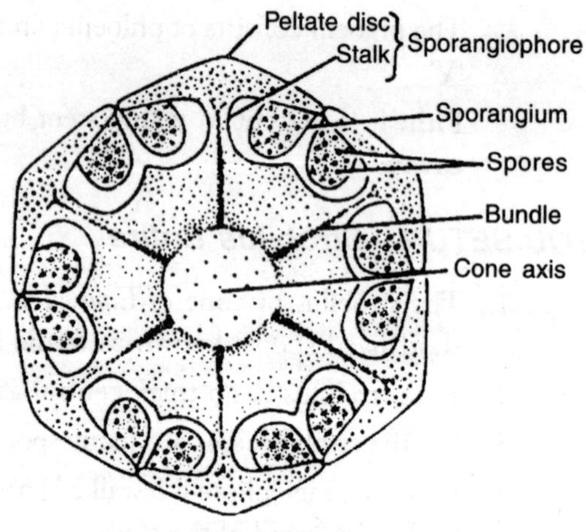

Fig. 4.13 *Equisetum.* T.S. cone

EQUISETUM—L.S. of the Strobilus

1. *Equisetum* is a homosporous plant. The L.S. of the cone shows a central axis, which bears sporangiophores that are placed at right angles to the central axis.
2. Each sporangiophore is an umbrella like structure with a peltate disc. On the underside of the disc 5–10 sporangia are borne as a ring around the stalk.
3. The mature sporangium is an elongate sac like body with a rounded apex. It is covered by a jacket, which is several cells in thickness of which the innermost cell layer serves as tapetum.

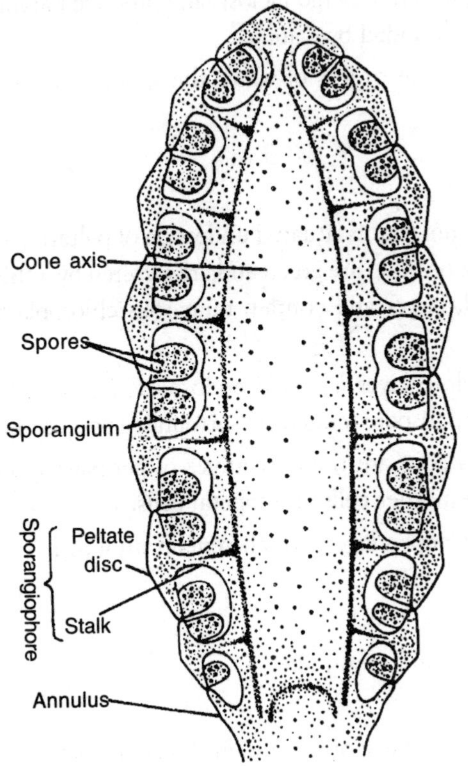

Fig. 4.14 *Equisetum.* L.S. cone

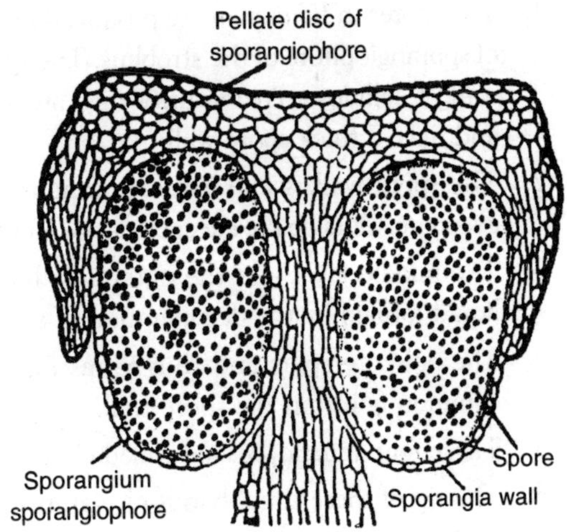

Fig. 4.15 L.S. of sporangiophore of *E. debile* showing two mature sporangia containing spores

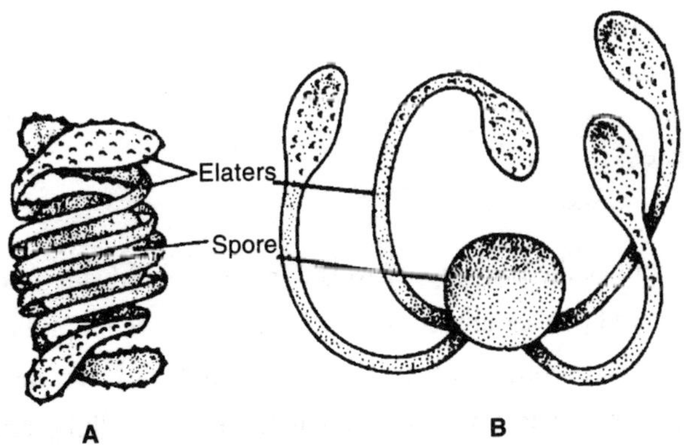

Fig. 4.16 *Equisetum.* Spores. **A.** Spore with elaters coiled, **B.** Spore with elaters uncoiled

4. The sporangium is filled with spore mother cells, which undergo meiosis and produce haploid spore tetrads. The spores are round, light and surrounded by elaters.
5. At maturity the strobilus axis elongates slightly. Due to loss of water the sporangiophores shrink and fall apart.

EQUISETUM—Spores and Elaters

1. The spores in *Equisetum* are produced in sporangium that is situated underside of peltate disc of sporangiophore of the stroblius. The spores when young are green and are covered by a thin wall of cellulose. At maturity they are spherical, larger and contain numerous chloroplasts and have a thick spore wall.
2. The spore wall is four layered. Innermost cellulose layer is called endospore, followed by exospore. Outside the exospore is the middle layer perispore (cuticular) surrounded by epispore.
3. The middle layer and the outermost layer of epispore are the derivatives of periplasmodial liquid formed by the degeneration of some spore mother cells and tapetal cells.
4. The epispore splits and thus forms four distinct appendages, which remain attached on a common point on the spore wall.
5. These four appendages are called elater.
6. The tip of these spiral bands or elaters is flattened and spoon like.
7. The elaters are hygroscopic in nature.
8. In moist conditions the elaters remain coiled round the spores while in dry conditions they separate and become free.
9. Elaters help in the dehiscence of the sporangia and dispersal of spores by parachute mechanism.

4.4 MARSILEA

Dvision — Pteridophyta
Sub-division — Pterophyta
Class — Leptosporangiatae
Order — Marsileales
Family — Marsileaceae
Genus — *Marsilea*

1. *Marsilea* is commonly called as 'pepper wort', and it is worldwide in distribution.
2. They grow submerged or partially out of water and they are amphibious in nature.
3. The plant body is a sporophyte, which is differentiated into rhizome, roots and leaves.
4. The younger parts are covered with hairs, which are multicellular and unbranched.
5. The rhizome is long, slender well branched grows on or below the soil surface and attains a length of 2.5 mts.

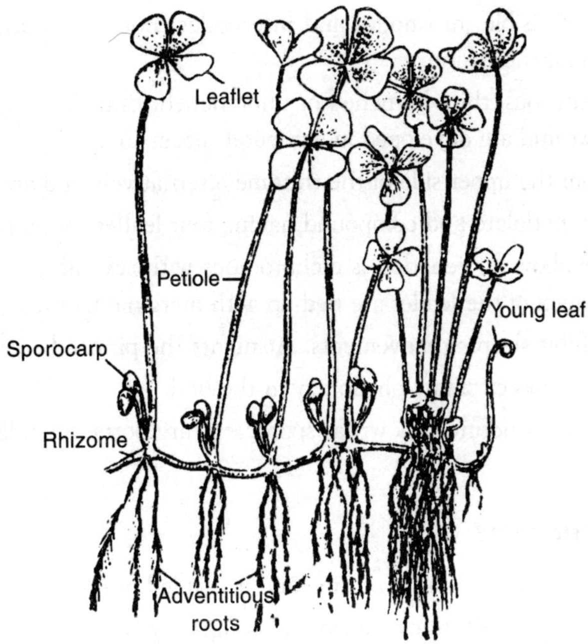

Fig. 4.17 Portion of sporophyte of *Marsilea minuta* showing habit

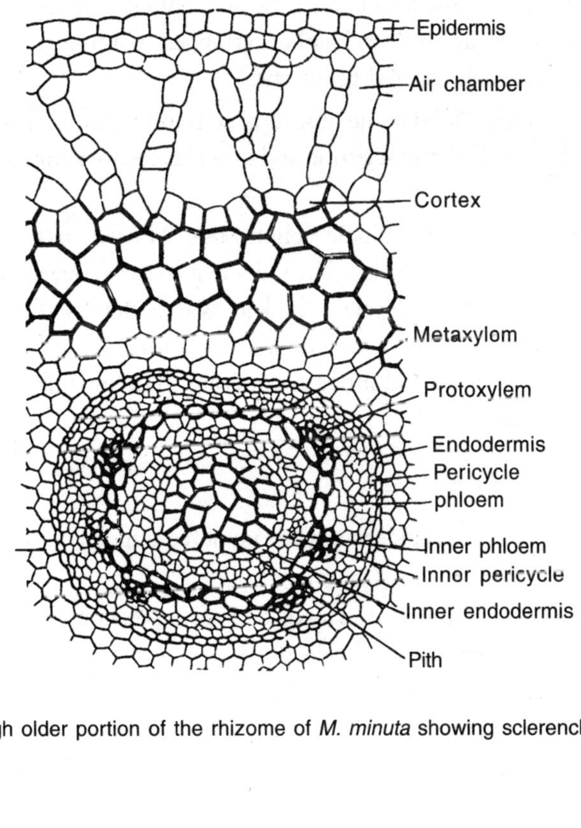

Fig. 4.18 T.S. through older portion of the rhizome of *M. minuta* showing sclerenchymatous pith

6. The rhizome is divisible into nodes and internodes. Both roots and leaves arise from the nodes in acropetal succession.
7. Roots are adventitious, thin, branched or unbranched. Branching is monopodial. The roots arise in two rows and are developed in acropetal succession.
8. Leaves arise from the upper side of the rhizome alternatively and are arranged in two rows.
9. Leaves are long, petiolate and compound having four leaflets or pinnae at its apex.
10. The leaflets are abovate, venation is dichotomous with several cross connections. The free veinlets at the apex of the leaflet are tied up with marginal loops.
11. The leaflets exhibit sleeping movements. At nights the pinnae become folded upwards.
12. The young leaf shows circinate phyllotaxy in the bud.
13. The spore bearing structures known as sporocarps are borne laterally near the base on the petiole.

MARSILEA—T.S. of Rhizome

1. The T.S. of rhizome is circular in outline and is divided into epidermis, cortex and stele.
2. Epidermis is single layered, thick walled and compactly arranged.
3. Cortex is differentiated into three regions, the outer cortex, middle cortex, and inner cortex. The outer cortex is parenchymatous and one to several celled thick. They are compactly arranged. A few tannin cells are present here and there.
4. The middle cortex consists of mainly the air cavities. This aerenchyma is an air storage tissue consisting of single layer of air chambers arranged in a ring. The chambers are separated from one another by single layered partitions.
5. The inner cortex is a solid tissue of several cells in thickness. The outer cell layers of the inner cortex are thick walled and sclerotic and inner layers are compactly arranged parenchymatous cells.
6. Stele is a medullated, amphiphloic siphonostele. The xylem is in the form of a ring surrounded by the phloem on both the sides. The phloem is in turn covered by the single layered pericycle on both the sides. The pericycle is flanked by endodermis. The disposition of various tissues from outside to inward is outer endodermis, outer pericycle, outer phloem, xylem, inner phloem, inner pericycle and inner endodermis. The medulla or pith is in the center and is sclerenchymatous.

MARSILEA—T.S. of Petiole

1. There is a single layered epidermis made up of rectangular cells. Beneath it is the layers of thin walled cells.
2. This is followed by middle cortex made cup of ring of air chambers. The inner cortex is a solid compact tissue which is several cells deep.
3. The parenchymatous cells are filled with starch and tannin.

4. The inner cortex is delimited by single layered endodermis.
5. The stele lies inside and is triangular in outline, has a single vascular bundle.
6. The xylem is V shaped with the opening towards the axis. Between the two arms of the xylem is the xylem parenchyma.
7. Surrounding the xylem is the phloem followed by the pericycle which is bounded by the endodermis.

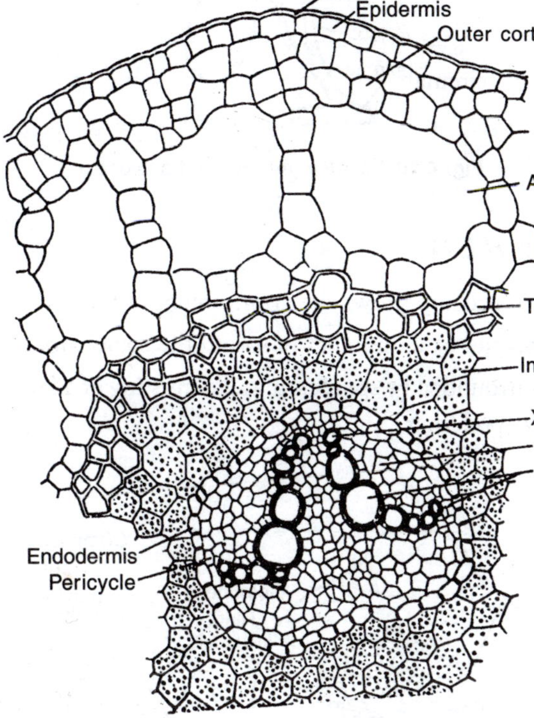

Fig. 4.19 T.S. petiole of *M. minuta*

MARSILEA—Sporocarp (Entire)

1. The sporocarps are special reproductive fruiting bodies.
2. They are borne at the base of the petiole either singly or in groups.
3. The sporocarps are bisporangiate (heterosporous).
4. Each sporocarp is a bean shaped or ovoid structure.
5. They are green and soft when young but turn black and hard nut like at maturity.
6. The place of attachment of peduncle with the body of the sporocarp is designated as raphe.
7. Just above the raphe the sporocarp shows two teeth like structures called tubercles or horns.
8. The sporocarp is bivalved (walled).

9. Each wall bears number of sori.
10. The sori of one wall (valve) alternate with the sori of other valve.
11. Each sorus bears both mega and microsporangia.

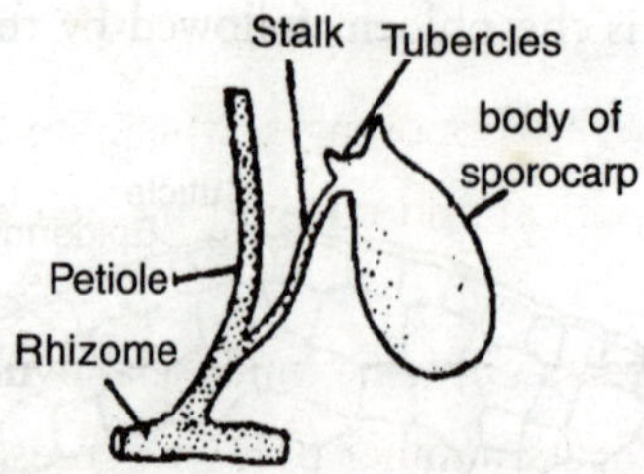

Fig. 4.20 *Marsilea.* An attached sporocarp

MARSILEA—Sporocarp H.L.S.

1. The H.L.S. of the sporocarp shows the arrangement of the sori in two alternating rows.
2. This section reveals the exact number of sori in the sporocarp.
3. The sporocarp is surrounded by sporocarp wall consisting of a layer of epidermis with stomata and two layers of hypodermis.

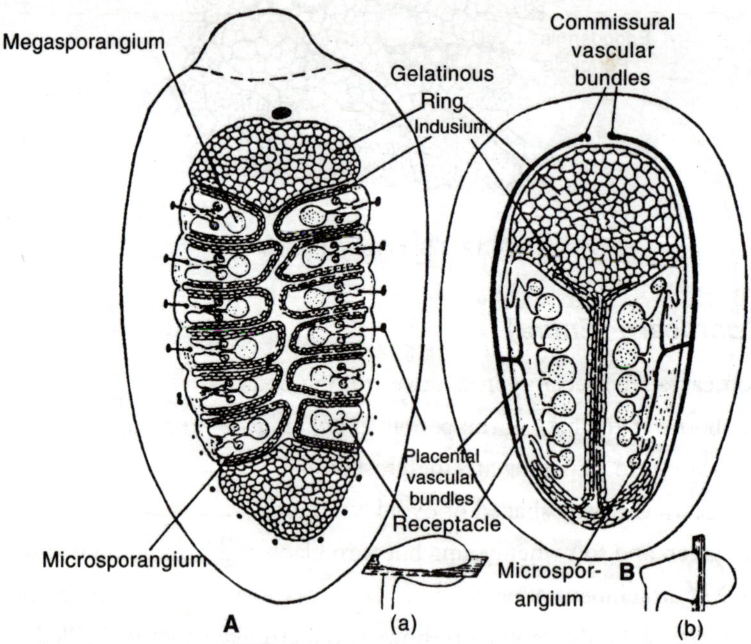

Fig. 4.21 Internal structure of sporocarp. **A.** Horizontal section through sporocarp. **B.** Vertical transverse section through sporocarp (*a* and *b* indicate planes of sections)

4. The megasporangia are seen at the apex of the receptacle and develop first.
5. The microsporangia are seen on the sides of the receptacle and develop later.
6. The sorus is thus gradated and sporangia develop in basipetalous manner.
7. The placental vascular bundles and cortical bundles are cut in transverse plane.
8. The gelatinous ring appears on either side.
9. A ring of indusium covers the sporangia, which are borne on the receptacle.
10. Each sporangium is spherical and covered by a jacket. The inner most layer of the wall is tapetum.
11. The megasporangium bears only one functional ovule.
12. The microsporangium bears number of microspores.

Index